Humanity 2.0

Also by Steve Fuller

SOCIAL EPISTEMOLOGY

PHILOSOPHY OF SCIENCE AND ITS DISCONTENTS

PHILOSOPHY, RHETORIC AND THE END OF KNOWLEDGE
The Coming of Science and Technology Studies (2nd Edition) (*co-authored*)

SCIENCE (CONCEPTS IN THE SOCIAL SCIENCES)

THE GOVERNANCE OF SCIENCE
Ideology and the Future of the Open Society

THOMAS KUHN
A Philosophical History for Our Times

KNOWLEDGE MANAGEMENT FOUNDATIONS

KUHN VS POPPER
The Struggle for the Soul of Science

THE INTELLECTUAL
The Positive Power of Negative Thinking

THE PHILOSOPHY OF SCIENCE AND TECHNOLOGY STUDIES

THE NEW SOCIOLOGICAL IMAGINATION

THE KNOWLEDGE BOOK
Key Concepts in Philosophy, Science and Culture

NEW FRONTIERS IN SCIENCE AND TECHNOLOGY STUDIES

SCIENCE VS RELIGION?
Intelligent Design and the Problem of Evolution

DISSENT OVER DESCENT
Intelligent Design's Challenge to Darwinism

THE SOCIOLOGY OF INTELLECTUAL LIFE
The Career of the Mind In and Around the Academy

SCIENCE (THE ART OF LIVING)

Humanity 2.0

What it Means to be Human Past, Present and Future

Steve Fuller
University of Warwick, UK

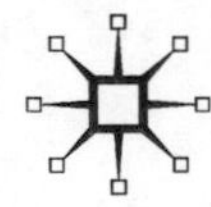

First published 2011 by
PALGRAVE MACMILLAN

Palgrave Macmillan in the UK is an imprint of Macmillan Publishers Limited, registered in England, company number 785998, of Houndmills, Basingstoke, Hampshire RG21 6XS.

Palgrave Macmillan in the US is a division of St Martin's Press LLC, 175 Fifth Avenue, New York, NY 10010.

Palgrave Macmillan is the global academic imprint of the above companies and has companies and representatives throughout the world.

Palgrave® and Macmillan® are registered trademarks in the United States, the United Kingdom, Europe and other countries

ISBN 978-0-230-23342-3 hardback
ISBN 978-0-230-23343-0 paperback

This book is printed on paper suitable for recycling and made from fully managed and sustained forest sources. Logging, pulping and manufacturing processes are expected to conform to the environmental regulations of the country of origin.

A catalogue record for this book is available from the British Library.

Library of Congress Cataloging-in-Publication Data

Fuller, Steve, 1959–
Humanity 2.0 : what it means to be human past, present and future / Steve Fuller.
p. cm.
Includes index.
ISBN 978-0-230-23343-0 (pbk.)
1. Social sciences–Moral and ethical aspects. 2. Social ethics. I. Title.
H61.F848 2011
301.01–dc23 2011016894

10 9 8 7 6 5 4 3 2 1
20 19 18 17 16 15 14 13 12 11

Printed and bound in Great Britain by
CPI Antony Rowe, Chippenham and Eastbourne

Contents

List of Tables

Introduction: What is Humanity 2.0?

This book is about *Humanity 2.0,* the emerging object of both social science and social policy in the 21st century. A good way to understand this new entity is to consider a deep ambiguity in what 'humane' means in perennial calls for the 'humane treatment of animals'. If it means that animals should be treated as humans are *normally* treated, then the charge is not very demanding. Aside from close friends and family members, we regard most people with an attitude of benign neglect, what Norman Geras (1998) rather precisely calls 'the contract of mutual indifference'. Such an attitude works well in a society of self-motivated individuals with reasonable resources at their disposal to achieve their ends, as envisaged by, say, Adam Smith. However, in a society lacking such initial endowments, all sorts of misery and deprivation would end up being tolerated as long as the victims remained out of sight. Notwithstanding significant pockets of welfare states and socialist regimes, today the world as a whole approximates such a state of benign neglect. If one takes this sociological fact seriously, then one may easily conclude that those who call for the humane treatment of animals are engaged in special pleading – in effect advocating *better* treatment for animals than humans, say, in terms of the scrupulousness with which suffering would be monitored and addressed.

Now of course, there is a more charitable spin to calls for the humane treatment of animals. It might be meant to recall levels of concern and care that we have yet to reach even in our own species. In that case, our treatment of animals is designed to hold up a mirror to our souls: We treat animals poorly because we treat humans poorly. While this argument may be rhetorically effective in gaining our attention, the lessons to be learned are far from clear. After all, a concerted policy effort to relieve all human suffering would be enormously time- and energy-intensive.

Moreover, it would increase the priority given to our own species *vis-à-vis* other species. Presumably this is not exactly what the animal advocates had in mind! On the contrary, they often speak as if the privileging of *Homo sapiens* itself is a major source of 'cruelty to animals', as the anti-humane approach has been traditionally called. Thus, theorists of 'animal liberation' have argued that the criteria for moral and perhaps even cognitive worth should lose their anthropocentric character (Singer 1975). Darwinian accounts of humans possessing an 'overdeveloped' cerebral cortex have added to this subtle denigration of distinctly human traits, so that the classic Kantian aspiration to universalise one's judgements has come to be supplanted by such animal virtues as, on the cognitive side, maximum adaptability and, on the moral side, pain avoidance. Unreconstructed humanists might be forgiven for thinking that these developments amount to a dumbing down of the *summum bonum,* or its modern equivalent, 'quality of life'.

One interesting and increasingly popular way to resolve our vexed sense of what it means to be human is to treat the possession of an animal body as only contingently related to our humanity. To be sure, this suggestion flies in the face of not only evolutionary biology but also most phenomenological philosophies and psychologies. Yet, at the same time, it does appeal to those philosophical and especially theological traditions in which human identity is tied primarily to 'consciousness', 'mind' or 'spirit', understood as a substance that is potentially subject to multiple material modes of conveyance, and hence 'resurrection', in the language of Christian eschatology (cf. Johnston 2009). In this argument, semi-siliconised cyborgs or outright computer androids might function equally well – if not more efficiently – as successor vehicles for the transmission and cultivation of what is distinctive about our being, whilst avoiding many if not all of the liabilities of human biology. It is perhaps no accident that the key theorists of the three most ambitious strategies of the past half-century to enable humans to transcend their biological moorings – cybernetics, artificial intelligence and prosthetic enhancement – have been all Unitarians, devotees of the tradition of Christian dissent that identifies human uniqueness with divinity itself. The theorists I have in mind are, respectively, Norbert Wiener (1964), Herbert Simon (1977) and Ray Kurzweil (1999). Together they present a quite different sense of 'Humanity 2.0' from those fixated on the 'humane' treatment of animals. They do not presume that the successor species of *Homo sapiens* will necessarily be a primate.

Whether a new and improved humanity ends up being resurrected through genetic mutation or digital programming, the stark difference in our future prospects points to 'Humanity 2.0' as a bipolar disorder that runs deep and long through Western culture. In more theologically inspired times, the distinction was cast as a future in which we either ascended into the heavens or were unceremoniously buried in the earth. The first wave of secularisation in 17th century Europe canonised the modern metaphysical expression of the divide as the 'mind-body problem'. Several other philosophically salient distinctions subsequently cut across this problem, most notably perhaps, individual-collective: Those who attributed minds primarily to individuals often rank- ordered individuals by the mental clarity (aka intelligence) that their bodies (aka brains) permitted, while those who attributed minds primarily to collectives tended to treat individuals as expendable means to an overarching spiritual end. In the 19th century, the biological and the psychosocial sciences began to define this distinction in terms of a 'typological' *vs.* a 'populational' conception of species (Mayr 1970). Now, early in the 21st century, following on rapid and convergent advances in the nano-, bio-, info- and cogno-technosciences, the mind-body problem is increasingly operationalised, if not outright replaced, by those who, on the one hand, would continue to anchor humanity in our carbon-based bodies or those who, on the other, would leverage humanity into more durable silicon-based containers.

The book before you aims to provide a comprehensive set of historical, philosophical and sociological resources to ask your own questions and draw your own conclusions about Humanity 2.0. It develops themes that have appeared in several of my recent books: Fuller (2006a: chap.5), Fuller (2006b), Fuller (2007a: chap. 6), Fuller (2007b: chap. 5) and Fuller (2008a). I have spent much of the past decade engaged in redefining the foundations of the social sciences in the face of a pincer attack from biology and theology. While I believe that Darwinism poses a much greater threat than Christianity or Islam to the future of humanity as a normatively salient category, none of these challengers can gainsay the genuine achievements that the social sciences have made over the past two centuries in extending the range of our existential horizons. Readers who wonder what I might have in mind could do worse than consult the now very unfashionable attempt begun in the late 1960s by the political scientist Karl Deutsch to construct a literal inventory of social science's achievements (Deutsch et al. 1986). I believe that this project deserves to be re-launched in the very near future.

Humanity 2.0 is divided into five chapters, each beginning with a paragraph that summarises the argument. The sequence proceeds by the following logic. Chapter 1 plants the reader *in medias res* by addressing the primordial mix of biology and ideology – or 'race and religion' – from which the social sciences emerged and through which they must continue to navigate. Here I am interested in the impact that a science of 'all and only humans' has had on how we treat each other in the present, the past and, most importantly, the future. The grounds for establishing a 'welfare state' is the red thread that runs through my considerations. Chapter 2 locates these developments in the deep history of philosophy, where the bipolar disorder alluded to above comes into sharp focus. I trace this to a schism already present in the High Middle Ages, the source of what I dub alternative 'mendicant modernities', by which I mean the difference between the more earth-bound Dominicans and the more heavenly oriented Franciscans, the two Christian mendicant orders that staffed the first universities. I end on the topic that takes up Chapter 3, namely, the ways in which new 'converging technologies' promise to transform the very constitution of the human species by allowing us to live longer, more productive lives. While the technologies now at our disposal may be revolutionary, the guiding ideas themselves are not – or at least not anymore, given their prominence in the positivist founding of the social sciences in the early 19th century. However, the focus of the chapter is on the challenges that the 'CT agenda' pose to how we think about public policy in the 21st century. These challenges are epitomised by increasing talk of 'transhumanism' and 'enhancing evolution'. Chapter 4 takes the animus behind these challenges to their theological roots, specifically the Abrahamic doctrine that humans are created 'in the image and likeness of God'. In recent years these roots have been rejuvenated by scientific creationism and intelligent design theory, versions of natural theology that refuse to accept the Neo-Darwinian orthodoxy in biology that would cast the difference between humans and other animals as merely a matter of degree, not kind. But what exactly would it mean to live 'in the image and likeness of God' in the 21st – not the 13th or 17th – century? Chapter 5 considers the prospects in terms of Humanity 2.0's moral horizons. Here I return to the philosophical basis for the welfare state, distributive justice, which, so to speak, recycles evil into good by imposing taxes on those who originally gained wealth by exploiting others. But of course, the principle may be carried to an extreme, according to which one needs to have practiced evil in order truly to do good – if only to experience the difference. I explore the shape that this radical idea, which sustained the faith of Thomas Malthus but not of Charles Darwin, might take in years to come. Crucial here is the prospect that people might be persuaded that 'death' does not mark an end to life but a translation into a medium that permits its continuation and enhancement.

My first word of thanks is to Nico Stehr, who, in cooperation with Liana Giorgi and Ronald Pohoryles at the Interdisciplinary Centre for Comparative Research in the Social Sciences in Vienna, spearheaded the European Union Sixth Framework Programme that enabled me from October 2006 to November 2007 to discuss with scientists and policymakers around the world about how the CT agenda is updating our conceptions of what it means to be human. They made it possible for me to spend 3.5 hours on the phone and in person interviewing Mihail Roco, the principal architect of the US initiative in this area. I also spoke at length with Ronald Kostoff, recently retired from the US Office of Naval Research, who for many years had been the in-house source of scientometric insight in US science policy. Among other interviewees and collaborators who influenced the shape of my thinking on this matter include Anders Sandberg (Oxford), Max Lu (Queensland), Yair Sharan (Tel-Aviv) and V.V. Krishna (Jawaharlal Nehru University). Let me also acknowledge the exceptional assistance of Albert Tzeng in amassing and analysing quantitative data relating to the CT agenda, even as it diverted him from completing his pioneering study of sociology's institutionalisation in East Asia. I had an opportunity to experience first hand CT's impact on US social science research at Arizona State University on 19–21 April 2007, courtesy of David Guston, Director of the Center for Nanotechnology in Society.

Among those who provided venues for airing earlier versions of the theses defended in these pages, special thanks go to Jan van Bouwel (Ghent), Noel Castree (Manchester), Piet Strydom (Cork), Tom Staley (Virginia Tech), Mark Porrovecchio (Oregon State), my Warwick colleagues Nickie Charles and Bob Carter, and last but not least, Vikram Seth, the Warwick graduate student responsible for staging the first TEDx lectures in Europe on 28 February 2009. Longer-term forms of gratitude go to Alf Bång and Christina Erneling of the Institute for Communication Studies, Lund University Campus Helsingborg, for hosting excellent international summer schools around the 'open society' theme over the last five years. The 2009 school featured an event centred on issues that form the basis of Chapter 4. Special thanks here to my interlocutors Ian Jarvie (York-Toronto), Jeremy Shearmur (Australian National University) and Inge-Bert Täljedal (Umeå). Among my other interlocutors have been Nicholas Maxwell (University College London),

Chris Renwick (York), Rachel Armstrong (University College London), and John Harris (Manchester). At Warwick on 19 May 2008, Harris and I debated the motion: 'There is no scientific basis to the concept of humanity'. Harris supported the motion; I opposed it. Peter Sloterdijk spoke from the audience. It was a starting gun moment in the race to define Humanity 2.0.

Finally, I would like to thank those most immediately responsible for getting the book into final form and ready for publication: Philippa Grand, Shirley Tan, Annabel Huxley and, last but not least, Luke Robert Mason.

1
Humanity Poised Between Biology and Ideology

To be human is to identify both an animal and an ideal. The clearest precedent for this duplex sense of humanity is the theological discourse surrounding the person of Jesus in Christianity as both 'Son of Man' and 'Son of God'. However, to begin here would be to wade in at the deep end of a problem that has repeatedly if diffusely vexed the social standing of science throughout the modern period. Indeed, if one had to identify two boundary issues for the 'social sciences' as a body of knowledge concerned with *all and only* human beings, they would be *race* and *religion* – that is, how we relate to our biological roots and our aspirations to transcend those roots. Clearly whatever we are as a matter of fact, many if not most of us are not satisfied with that as the final word. The first chapter takes this dissonance head on. I begin by laying down the main historico-philosophical markers, which still inform contemporary debates about what it means to be human. Sections 2–5 deal with how these debates have defined the directions that sociology and social policy have taken in the 20th century. Most notable is the centrality of the welfare state as a battleground for constituting humanity as a biological *vis-à-vis* an ideological entity, various resolutions of which may be found across the European continent, not least in the 'national socialism' of Germany and Scandinavia. Against this history, the United Kingdom stands out for its institutional segregation of the social and the biological sciences, despite that nation's pioneering empirical investigations in both fields of study. The establishment of the UK's first sociology chair at the London School of Economics in 1907 provides a convenient hook for exploring this point. Sections 6–8 demonstrate how alternative ideological conceptions of humanity continue to divide our attachment to our biological natures. On the one hand is the identity politics of embodiment associated with the New Left; on the other, the more exotic, even non-carbon embodiments associated with the search for extraterrestrial life.

1 Science's twin taboos: Race and religion

In October 2007, the Nobel laureate and co-discoverer of the double helix structure of DNA, James D. Watson, was hounded out of the UK during a book tour for comments he made to the effect that Western development aid to Africa was wasted because of the relatively low intelligence of its recipients (Milmo 2007). He was subsequently forced to resign from the most venerable US biomedical research facility, Cold Spring Harbour Laboratory in New York, which he had directed for 35 years of its most significant growth, after it had been the site of the Eugenics Record Office in the first half of the 20th century. Almost exactly two years earlier, Michael Behe, a tenured professor of biochemistry at Lehigh University in Pennsylvania, had testified under oath in the Third US Circuit Court that the scientifically credentialed form of creationism, known as 'intelligent design', deserved a place in the public high school science curriculum alongside Neo-Darwinian evolution as an explanation for the origin of life. Not only did Behe's side lose the case but also Behe himself has been subsequently subjected to personal vilification, abetted by his department's official – and continuing – dissociation from his views (LUDBS 2007).

These two cases touch on science's twin taboos: race and religion. It seems that science, especially biological science, cannot live with – or without – them. Together they define the limits of respectable public scientific discourse. To be sure, race and religion breach scientific respectability from opposite directions: Whereas racism makes a fetish out of the persistent diversity of the human population, creationism overplays the significance of our common descent from a deity in whose 'image and likeness' we are supposedly created. Thus, racists and creationists propose alternative utopian visions for humanity: the former project an ideal world of well-bounded limited populations in ecological equilibrium, while the latter envisage that our ever expanding and more mobile numbers will permanently transform the planet for our collective benefit. Their respective visions of history – as retold in, on the one hand, Weikart (2005) and Pichot (2009) and, on the other, Passmore (1970) and Noble (1997) – may be captured in a single dialectical phrase: *evolution vs. progress*.

There used to be an entire science dedicated to debating creationism and racism on empirical grounds. It was called 'anthropology', named after the title of a 1798 work by the German philosopher Immanuel Kant (1724–1804), which was dedicated to how – and whether – the different races might embody the same Enlightenment ideal of a *Weltbürger*; a 'world-citizen' (Kant 1798). For much of the 18th and 19th centuries, the two positions travelled under the epistemologically sanitised labels of *monogenesis* and *polygenesis*, respectively (Harris 1968). But their proponents did not quite match up to today's creationists and racists: There were religious and secular thinkers on both sides of the divide. On the one hand, the party of monogenesis was composed of New Testament promoters of the 'universal brotherhood of man' and Enlightenment optimists pursuing human perfectibility, such as the French philosopher Nicolas de Caritat, aka Marquis de Condorcet (1743–1794). On the other hand, the party of polygenesis consisted of literal adherents to the multiple dispersals of human life postulated in the Old Testament as well as Enlightenment sceptics including the Scottish philosopher David Hume (1711–1776) who regarded 'humanity' as the brand name for a variety of upright apes.

The difference between monogenesis and polygenesis is epitomised in a question: *Are the variety of beings that pass for humans the result of one or multiple origins?* I say 'pass' because at the time few denied, say, the *prima facie* grounds for moral concern about the hereditary enslavement of Africans in Europe and the Americas. However, that shared concern did not necessarily translate into a belief that Africans and Europeans shared a common ancestry, or at least one that was sufficiently strong to overcome their differences in appearance and mode of being.

The mindset of today's animal rights activists provides a point of reference to these debates about slavery. While the activists strongly object to the suffering endured by caged laboratory animals, most would stop short of according them civil rights because they doubt the animals' competence to take full responsibility for their actions. (Interestingly this position is a climb down from that of the original Animal Liberation Front of the 1970s, which routinely made provocative associations between caged animals and enslaved humans.) Similarly, it is one thing to justify the emancipation of slaves in terms of upholding universal human rights *à la* monogenesis, and quite another in terms of supporting the cultivation of life under conditions where it is likely to flourish *à la* polygenesis. Do slaves suffer and revolt out of their God-given sense of natural liberty, which is shared by all humans, or out of their instinctive rejection of unnatural living conditions, which vary across animals?

The dispute between monogenesis and polygenesis gradually subsided as people eventually accepted the plausibility of a negotiated settlement, which was brought into effect by Charles Darwin (1809–1882) with the publication of *On the Origin of Species* (1859). Human races are environmentally reinforced genetic subdivisions, or 'sub-species', which descended from a common hominid ancestor, which itself descended from ancestors common to other species, which ultimately goes back to a primordial soup out of which life on earth first came into existence. On the one hand, Darwin did not explicitly associate life's emergence from the primordial soup with God's creative efforts; the process could have been equally the result of a divine spark and of an entirely self-organising process. On the other hand, Darwin not only refused to rank human races, he even stopped short of admitting that humans were the noblest species. Indeed, Darwin's *The Descent of Man* famously ends with his declaration that he is happier knowing that he descended from baboons – not Caucasoids – than the fierce inhabitants of Tierra del Fuego he encountered on the *Beagle* (Darwin 1871). If anything, Darwin's studied anti-racism looks like the sort of 'species egalitarianism' nowadays associated with the animal liberation proponent Peter Singer (1999). Darwin's rhetorical genius lay in refusing to take a clear stand on the matters that divided the creationists and racists of his day, and hence leaving the nature of humanity profoundly ambiguous. However, 150 years after the publication of *The Origin of Species*, Darwinian diplomacy appears to be unravelling with the resurgence of both racism and creationism as potentially scientific propositions.

Anyone familiar with American legal history will be struck by the similarity between the rhetoric now used to 'separate' religion from science and that introduced a century ago to 'segregate' Blacks from Whites. In the case of race, the precedent was set in the generation following the abolition of slavery by the US Supreme Court in *Plessy vs. Ferguson* (1896), a decision that was eventually overturned with *Brown vs. Board of Education* (1954). In *Plessy*, the justices ruled that formal recognition of racial equality did not require that the races be given access to common facilities. The justices appeared to believe that racial equality was compatible with a caste system that restricts the mutual access of Blacks and Whites. At the same time, they also stressed that the provision of separate schools, washrooms or rail coaches for Blacks and Whites did not *ipso facto* imply that Blacks would receive inferior facilities. This even led some Whites to complain that *Plessy* compelled the construction of facilities for Blacks that might go underutilised.

A similar segregationism *vis-à-vis* religion was explicitly made in the Pennsylvania case mentioned earlier in this article: *Kitzmiller vs. Dover Area School District* (2005). Here, circuit court judge John E. Jones III expressly refused to pass judgement on intelligent design's truth, only on its status as science (US District Court 2005). Indeed, he suggested that intelligent design may be true in some *other* sense that might be taught outside the science class. Ironically, this line of argument – which is sometimes called the 'double truth' doctrine after the medieval scholastics – was the one used by the Roman Catholic Church against Galileo Galilei (1564–1642) to limit the reach of his scientific claims against theological interpretations of the Bible. After all, Catholic missionaries in China were promoting science as one of the fruits of their religion, while refusing to have it impinge on religion at home. Now, with *Kitzmiller*, the tables seem to have been turned. Just as the Church was happy to let Galileo conduct his research if he stopped promoting it as superior to Catholic doctrine, intelligent design could be taught as one wished but not as science (Fuller 2008a: chap. 2).

Despite the similarities, there is an obvious difference between the legal fates of racial and religious segregation in the USA: Discrimination on the basis of race has been overcome, while discrimination on the basis of religion has been intensified. But in both cases, the consequences have been perverse for science, and resulted in the coinage of mildly euphemistic expressions such as 'genetic diversity' and 'intelligent design' to keep the issues represented by race and religion in the scientific debate. Interestingly, both strategies are compelled by virtue of traumatic historical events that have inhibited any role that race or religion might ever again play in science. I shall illustrate the point first with race, followed by religion.

During their lifetimes and until the end of World War II, Charles Darwin and Herbert Spencer (1820–1903) were regarded as the main promoters of the theory of evolution by natural selection, as popularised in Spencer's expression, 'survival of the fittest'. Both were, broadly speaking, *laissez faire* liberals, sceptical of the role that states might play in reversing natural dynamics. Confessing ignorance of the mechanisms of heredity and inclined to believe that natural selection would always ultimately trump artificial selection, both refrained from endorsing the original version of eugenics touted by Darwin's cousin, Francis Galton (1822–1911).

Of course, this did not stop 20th century developments in genetics, including their eugenicist applications, from being treated as extensions of Darwin's and Spencer's work. Back then, the difference between the two was seen to lie more in emphasis than substance: Spencer focussed on the implications of evolution for contemporary human concerns, whereas Darwin generally avoided any such talk. This made Spencer the most influential spokesperson for evolution in the final quarter of the 19th century, even including Darwin's public defender, Thomas Henry Huxley (1825–1895).

However, all of that changed with the rise and fall of the Nazis and the atrocities they carried out in the name of eugenics, which led to the subsequent stigmatisation of 'survival of the fittest' policies. The postwar political climate was such that evolutionary theory was potentially held liable for Hitler's carnage. The diplomatic solution, again to Darwin's advantage, was to jettison Spencer as a 'Social Darwinist' (Hofstadter 1944). The phrase had neutrally referred to the extension of Darwin's ideas to human affairs, typically on the basis of what he himself had provided in *The Descent of Man*. But thereafter 'Social Darwinist' referred quite specifically to the overextension, and hence misuse, of those ideas. Spencer was the obvious target of this semantic shift, as he had been Darwin's most visible promoter in the social sciences and politics, not least in Germany. Darwin's studied silence on human affairs left him as the only 'politically correct' 19th century ancestor to the postwar synthesis forged between natural history and Mendelian genetics. However, the conceptual cost of overcoming racism in this fashion was excluding by default any deep studies on *Homo sapiens* from modern evolutionary theory.

This point is epitomised in UNESCO's influential 1950 statement on 'the race question' (Brattain 2007). Asserting the biological unity of humanity, it portrayed claims to racial difference as little more than socially-based 'ethnic' stereotypes ultimately grounded in unscientific prejudices. The coalition of distinguished social and natural scientists who were involved in finalising the statement ensured its legitimacy. Nevertheless, it appeared shortly before DNA-driven breakthroughs in molecular biology revolutionised our understanding of genetics by providing a more fine-grained sense of both what unifies and differentiates humanity. Two large-scale projects from the past quarter-century, one devoted to sequencing the common human genome and the other to charting the course of human genetic diversity, represent the fruits of that revolutionary endeavour. It is the latter that concerns us here.

The human genetic diversity project draws on an eclectic mix of genetics, archaeology and linguistics to follow the migration patterns of the Earth's peoples (Cavalli-Sforza 2000). Courtesy of IBM and *National Geographic* magazine, the project acquired a populist dimension in 2005, when it invited people interested in tracing their family histories to contribute personal information online in exchange for access to the project's collective knowledge base. So far, more than a quarter-million people worldwide have complied. However, the United Nations' Permanent Forum on Indigenous Issues has recommended suspending the project because its results might be used opportunistically either to undermine affirmative action policies – if the disadvantaged indigenes turn out to be genetically similar to the dominant foreigners – or, in keeping with earlier racist policies, to 'repatriate' those who happen to have a large number of medically relevant alleles belonging to peoples normally resident elsewhere (Harmon 2006).

No doubt the very idea of genetic diversity taps into the heritage of *Rassenwissenschaft* ('race science') and *Rassenhygiene* ('racial hygiene'), an original stronghold of Darwinism and perhaps the most exciting field of German biomedical science prior to Hitler's ascendancy (Proctor 1988). Its intellectual leader Alfred Ploetz (1860–1940) promised no less than perpetual peace if the Earth's peoples confined themselves to lands suitable to their respective genetic make-ups. For this he was nominated for a Nobel Prize – albeit in 1936, late in life and after the Nazis had come to power. A staunch anti-imperialist, Ploetz had called for a massive resettlement policy of the multiple ethnic groups of the newly amalgamated Second Reich as early as 1895, with each 'homeland' urged to design its own indigenous social security system, customised to the specific health needs of its people.

Originally, this proposal was made in the spirit of social democracy, even socialism, which by the 1930s had turned the Nordic countries into welfare states (Broberg and Roll-Hansen 1997). However, it took a sinister turn in the wake of Germany's defeat in World War I in 1918, when it became incorporated into the campaign platform of the newly formed National Socialist Party (Fuller 2007b: chap. 5). In contrast, the human genetic diversity project is operating in a political climate less prone to totalitarian abuse. An interesting witness here is the German-born and trained Harvard evolutionist, Ernst Mayr (1904–2005), who remained active until his death. Without ever endorsing Nazism, he never failed to assert the relevance of biologically grounded racial differences for medical and perhaps even legal purposes (Mayr 2002). Indeed, biomedical research provides growing evidence that race can indeed be used as a marker to influence physicians' diagnostics and recommended treatments and for preventive measures (Weigmann 2006; Kahn 2004). Yet, as long as 'politically correct' intuitions remain firmly anchored in the sentiments expressed in the original UNESCO document, there will be formidable barriers to allowing genetic diversity to explicitly inform policymaking.

As for religion, a US-style exclusion of religion from science is inscribed in the constitutions of many modern states. Usually the difference between religion and science is reduced to the distinction between private and public knowledge. Moreover, in the USA and elsewhere, the legal separation of church and state has evolved from preventing a specific church's dominance in civil society to preventing religion's influence altogether. The anchoring trauma here is the social discrimination originally suffered by wealthy, well-educated Christians in 17th century England who happened not to be members of the established church. They were compelled to start their own society, which over the next 150 years became the USA. The founders of this new nation resolved that never again would the same mistake be made: since 1791, the separation of state and church is derived from the first amendment to the US Constitution which states that 'Congress shall make no law respecting an establishment of religion, or prohibiting the free exercise thereof ...'.

Indeed, the separation of state and church has been pursued in the USA with a bloody-mindedness that now overlooks religion's distinctly positive impact on the development of science. It is a point that the original American settlers, as Puritan promoters of the Scientific Revolution, would have been the first to admit (Merton 1970). The expression 'intelligent design' taps into that founding sentiment by recalling the strong analogy that the 17th century scientific revolutionaries, most notably Isaac Newton (1642–1727), drew between the machine-making capacities of humans and the creative agency of God. In effect, to see life as the product of intelligent design is to conceive of biology as divine technology. This eventually led the US founding fathers to conceptualise the 'mechanism of government' as literally a second creation (Cohen 1995).

To be sure, intelligent design always had a heretical cast. It implied that God's power exceeds human power only by degree not kind. Instead of upright apes, we are demigods. This explains the initial plausibility of what is now a quite ordinary feature of modern scientific reasoning, namely, model-building. After all, it is one thing to design a machine that works on its own terms but quite another to think that the machine captures properties of the natural world to such an extent that it may be used as a basis for prediction and control. The scientific method honours that distinction as the difference between the 'reliability' and 'validity' in laboratory experiments and computer simulations. Thus, Michael Behe (1996) may be wrong in claiming that the cell is 'irreducibly complex' in the same sense as a mousetrap, all of whose parts must be in place to work. But to think that a cell might work like a mousetrap is very much in the spirit of the mechanistic worldview that launched modern science.

More generally, intelligent design theory taps into the vast majority of science that has been done under the assumption that nature is a unified rational whole; and humans have been specially created to understand, manage and possibly improve it, if not to bring it to outright completion. The philosophical term of art for this quality of nature is 'intelligibility' (Dear 2006). The assumption of intelligibility is shared not only by so-called young earth creationists, who claim on biblical grounds that the planet is only 6000 years old, but also physicists who continue to search for a grand unifying theory, and biologists, who seek a progressive direction to evolution. Darwin stands out in the history of biological science from both his great predecessors – Carl Linnaeus (1707–1778), Georges Cuvier (1769–1832) and Jean-Baptiste Lamarck (1744–1829) – and his great successors – Gregor Mendel (1822–1884), Sewall Wright (1889–1988), Ronald Fisher (1890–1962) and Theodosius Dobzhansky (1900–1975) – in his failure to find his faith in God bolstered by his research, though his many decades of intellectual labour were originally motivated by a search for intelligent design in nature.

Given the recent strong public expression of atheism (Dawkins 2006), the following question looms large: Can the degree of human cognitive privilege implied in the idea of intelligent design be denied without undercutting the basis for the most inspiring theoretical projects in science? Atheists, of course, say yes. However, in the *Critique of Pure Reason*, the cornerstone of modern Western philosophy, Kant answered with a resounding no, while quickly adding that just because we need to postulate the existence of God to justify the pursuit of science, it does not follow that God actually exists (Kant 1781). Kant had in mind Newton, whose exemplariness lay not only in the detail in which he worked out material motion in the known cosmos, but also in developing an artificial language – physics – that could lay reasonable claim to represent the divine standpoint, shorn of the partial subjectivity of his creatures.

However, Newton took a beating with the early 20th century revolutions in relativity and quantum mechanics, which empirically undermined some of his fundamental conceptual assumptions. These developments, combined with the perverse uses to which the physical sciences were put in the two world wars, shook the faith of many in nature's intelligibility. Even those scientists who continued to believe in God tended to conceive of the deity in more remote, sometimes irrational, terms than Newton or Kant would have deemed appropriate.

But would atheism do as well as a background belief for science? Hardly. Atheism's track record is limited to dispelling superstition and challenging dogmatism: It does not extend to promoting science (Fuller 2010: chap. 6). Consider the most intellectually substantial and attractive figure reasonably counted as an atheist – as opposed to simply a heretic or deist: David Hume. His famed scepticism cut against not only theologians, who saw nature's design as evidence for God's existence, but also scientists, who followed Newton's lead in thinking that they were on the verge of fathoming nature's inner workings (Schliesser 2008). Hume's counsel was ultimately a therapeutic one, later echoed by that other icon of Anglophone analytic philosophy, Ludwig Wittgenstein (1889–1951): We should lower our epistemic expectations and let go of the idea that some overall mastery of nature is to be had by either philosophical or scientific means.

While such advice might put worried minds at ease, it fails to explain the success of those who failed to heed it. After all, notwithstanding the postwar taboo on race, the revolution in molecular biology managed to bring genetics to the brink of bioengineering by the 1960s (Morange 1998). And in light of the ongoing success of both the human genome and the human genetic diversity projects, we might need to revisit eugenics with a more positive frame of mind to social experimentation. In the end, the question that continues to dog us is who exactly are 'we', the subject of this grand narrative that would unify our racial and religions differences. Once we get some agreement on that question, race and religion will cease to be taboo subjects.

2 Sociology's official anti-biologism and unofficial soft racism

When the social sciences are presented as the most progressive of the three main bodies of knowledge – that is *vis-à-vis* the humanities and the natural sciences – a story is told whereby the social sciences provide voice and direction for what the 18th century Enlightenment philosophers had called the 'project of humanity'. On the one hand, the social sciences incorporated the non-elite members of *Homo sapiens* whose lives did not leave the sorts of traces that 19th century humanists had deemed worthy of study and, on the other, they bore secular witness to old theological ideas that humans stand out from the rest of nature by virtue of their uniquely 'meaningful' activities. Moreover, the other two bodies of knowledge are presented as offering an unholy alliance of the theoretical and the instrumental. Illustrative of the humanities is the educational regime of philosopher-kings in Plato's *Republic*, in which a lifetime of contemplation becomes preparation for manipulative, authoritarian rule. In the case of the natural sciences, consider the application of the mathematical abstractions of Newtonian and Einsteinian physics to the exploitation of the earth and possibly beyond. In contrast, the social sciences appeared as the prime vehicle of humanity's self-realisation, treading a middle ground between the excesses of purely theoretical and purely instrumental knowledge, what Habermas used to – and critical realists still do – call the 'emancipatory' interest in knowledge (Bhaskar 1986).

This progressive image of the social sciences was shared by sociology's founding trinity: Karl Marx, Emile Durkheim and Max Weber. Although they envisaged 'humanity's self-realisation' in somewhat different political terms, these were clearly democratic and fell within the centre-to-left range of the late 19th century ideological spectrum. On this portrayal, which still has many attractive features, the social sciences perform a kind of epistemological chemistry, purifying the two earlier forms of knowledge – the humanities and the natural sciences – and combining their extracted essences into a stable higher form (a 'synthesis' to German idealists) that will allow humanity to become all it can be, both individually and collectively. This is the spirit in which one should understand the many chequered attempts to bring 'rational' order to society over the past two centuries, perhaps starting with the post-Napoleonic plan by Auguste Comte's mentor, Henri de Saint-Simon, to 're-organise' Europe but also including the spread of civil services, social welfare schemes and military conscription across entities called 'states', for which 'society' became the politically neutral face (Hayek 1952: chaps. 12–16). Thus, geographical boundaries came to carry teleological weight of potentially serious political import on the global stage: What is protected within one's borders is a way of life that is entitled to uninterrupted development.

The very idea that society can be 're-organised' implies that society can be treated as an organism, albeit an artificially designed one. Yet, after Herbert Spencer and his fellow-travelling Social Darwinists and Neo-Lamarckians tried to assign name 'sociology' to the discipline in charge of evaluating the biological implications of this idea, the discipline spent most of the 20th century studiously avoiding any clear position on the varied and changing character of biological knowledge. Sociology's 'holy trinity' provide three traditions of justifying this avoidance. Marx, who stayed truest to the Enlightenment project of humanity, believed that biologically-based differences, as reproduced through the limited social mobility permitted in class-based societies, constitute the main barrier to a truly egalitarian socialist regime. Durkheim, who was more concerned with carving out space for sociology as an academic discipline, pointed to the irreducibility of distinctively human endeavours to matters of individual survival and self-interest. As for Weber, his anti-biologism related to his training in law and political economy, which led him to define sociology in terms of the conventions and contracts formed by humans in historical time, without importing invariably speculative and possibly mythical notions of how these have been overdetermined by, say, heritable racial memories (Fuller 2006b: chap. 7).

Thus, many grounds were offered by sociology's founding fathers for enforcing a strong disciplinary boundary between sociology and biology. Weber's methodological grounds were blended with Durkheim's ontological ones and Marx's political ones into an updated formulation in Anthony Giddens' (1976) *New Rules for the Sociological Method*, which probably still captures the default position among workaday sociologists today. As we shall see in section four, over a hundred years ago, even before the combined influence of sociology's holy trinity was felt, and despite having been the home of Malthus, Darwin, Spencer, Galton and Huxley, Britain generated its own home grown aversion to sociobiologism. This is not to say that sociology ignored biology altogether. Rather, it was always implicitly present and almost always negatively marked – a residue of our primitive, even animal past in which kinship ties dominated over contractual arrangements and a default sense of 'ascribed' (as opposed to 'achieved') status and 'traditional' (as opposed to 'modern') thought prevailed. For example, it was clear that Francis Galton saw eugenics as a scientific solution to the lingering drag that ancestral habits of thought and action continued to exert on human social evolution. Indeed, he understood his project as much more applied sociology than applied biology (Renwick 2012: chap. 2). We would now call it a theory of societal reproduction – or, after Giddens (1984), 'structuration' – designed to create and maintain distance from our evolutionary past.

But beyond the appeal to inheritance as a lever for potentially radical social reform, biology also remained in a certain 'soft' conception of 'race' that continues to pervade public discourse about who does and does not belong in a given society. To illustrate the point, take two people who are charged with anti-social behaviour, an Asian who is the second-generation of his family to have been born in the UK and a White whose parents migrated from the United States. In the latter case, whatever other issues might be raised, one can be sure that they will *not* include 'unsuitability' of Americans to the British way of life. I call the conception 'race' because, especially when dealing with deviant behaviour, people's social adaptability is evaluated by their family origins. However, I call this racism 'soft' because it is based on a conception of heredity that is rarely documented for more than a few generations in which case the political question turns on the political and economic feasibility of tolerating the anticipated period of transition to normalcy. Doubts along these lines have been regularly mined by Fascist and anti-immigration parties. But far from being ignorant myth, soft racism taps into deep, scientifically unresolved issues concerning socio-biological causation. In the next section, these issues are projected onto the history of the welfare state, arguably the great policy success story of the social sciences, yet one achieved by taking some hard decisions on biology's relevance to social life.

The folk sociology of soft racism suggests that the various races *can* ultimately fit in, though some races may require more time than others, depending on how quickly they acquire the right susceptibilities to the new environment. Here I deliberately allude to Jean-Baptiste Lamarck's theory of the inheritance of acquired traits, the original theory of evolution, whose purposeful and progressive character distinguishes it most clearly from Darwin's. To be sure, biologists nowadays like to speak instead of 'horizontal gene transfer' to refer to the capacity of an alien body – be it introduced naturally as a microbe or artificially via xenotransplantation – to alter permanently an organism's genetic composition (Woese 2004). But all that is really conceded by that turn of phrase is that the acquisition of new traits need not be the product of the organism's conscious striving, as Lamarck had thought (Por 2006). Rather, the adaptiveness of one's offspring to an environment may be unwittingly enhanced simply through one's own regular exposure to that environment, if not the deliberate intervention of a more intelligent organism.

Whether evolutionary fitness is improved by design or by accident, soft racism remains an attractive explanatory framework as long as Michael Polanyi's (1957) slogan for tacit knowledge continues to hold intuitive appeal – to wit, that we always know more than we can tell. In that case, one is never explicitly taught how to be 'native' or 'normal' in a given environment but the relevant competence is somehow acquired in one's lifetime – perhaps by some folk sociological process of 'osmosis' – that improves the next generation's adaptive capacity. Biological research in this area, called *epigenetics*, has increasingly focussed on the hypothesis that even if the components of an offspring's genome are not altered by its parents' activities, how the components work together to express the offspring's specific traits may well be. This hypothesis potentially has many more physically invasive policy consequences than the more strictly Darwinian views associated with, say, the 'Baldwin Effect' (Weber and Depew 2003) or the 'extended phenotype' (Dawkins 1982), according to which organisms reconstruct their life-worlds in ways that benefit offspring with similarly expressed genes. In these cases, there is no suggestion that the reconstructed environments alter the process of gene expression itself, simply which sets of expressed genes are likely to be favoured.

The problem facing the epigenetic hypothesis is that while the relevant changes may be easy to detect in relatively simple organisms, they are difficult to establish in humans, given the longevity of each generation and hence the opportunities to acquire the relevant changes without altering the genome. Nevertheless, if one presumes a unity to biological nature, then it is likely that some version of the epigenetic processes that are observable in successive generations of insects and mice under laboratory conditions happens to humans under normal social conditions. The remaining question is whether this subtle form of soft racism is sufficiently manipulable in policy-relevant terms to revive a eugenicist agenda (Hunter 2008). Whether one treats this development as an updated version of 'Social Darwinism' (cf. Dickens 2000) or 'Neo-Lamarckianism' depends on the degree of genetic plasticity attributed to the organism and the degree of control attributed to its would-be manipulator: the more of both, the more Lamarckian in appearance.

To be sure, soft racism runs counter to what for over the last hundred years professional biologists have officially recognised as the 'Weismann Barrier', named for August Weismann, arguably the greatest German biologist of the 19th century (Mayr 1982: chap. 16). The honour is certainly apt if one thinks of biology in politically correct 20th century terms as a discipline strongly bounded from sociology, as indeed the barrier bearing his name implies. (However, if one includes biologists with a more 'open-borders' approach to the two fields, then pathologist Rudolf Virchow and embryologist Ernst Haeckel, who clashed publicly over the teaching of Darwin's theory, would certainly give Weismann a run for his money. Each in his own way contributed to the racial hygiene movement discussed in the next section.) The Weismann Barrier claims that we always inherit some combination of our parents' genes but never the changes that their bodies (including their brains) have undergone prior to our conception. Nowadays biologists call this discrete mode of inheritance 'vertical gene transfer', which stresses the idea that genetic information is conveyed exclusively through lines of familial descent.

To accept the Weismann Barrier is not to deny that genes change over time. However, these changes are conceptualised, in the first instance, as 'mutations', which is to say, by-products of the normal process of genetic transfer. A good way to think of these by-products is as slightly imperfect reproductions that can become significant over time if they survive and accumulate. That, in turn, depends on natural selection, which is understood as an independent process that, so to speak, 'blind tests' the fitness of successive generations to the environment. Indeed, the establishment of the Weismann Barrier is normally cited as marking the success of Darwin's over Lamarck's theory of evolution. Nevertheless, Darwin himself was in many respects still a pre-Weismannian thinker – and not simply because he died in 1882, a decade before Weismann proposed the barrier. Darwin's residual Lamarckianism also led him to conclude *inter alia* that women had no need for education once they had borne children because it would serve no genetically useful purpose. However, as suggested above, the Weismann Barrier is now under attack by the increasing importance attached to horizontal gene transfer, which reflects more than anything else our increasingly exact ability both to cause and to register lasting changes in organisms.

3 Alternative biological foundations of the modern welfare state: Germany and Scandinavia

Perhaps the most vivid example of the insecurity of the *cordon sanitaire* protecting sociology from biology in the 20th century pertains to the conflicting biological presuppositions that underwrote the foundation of the modern welfare state, arguably the most enduring achievement of social science-led policymaking on a grand scale. In retrospect, what is perhaps most striking about the advancement of the welfare state in the preceding century is the (natural-*cum*-social) science-based confidence of democratically elected politicians in engaging in unprecedented forms of coercion, ranging from relatively subtle ones like redistributive taxation to quite gross ones like physical displacement, mutilation and extermination. Armed with this brute fact, more sophisticated right-wing

US political commentators have tried to cast the overall history of the welfare state as one long march of 'liberal fascism' (Goldberg 2007), a tendency that has only increased in the wake of Barack Obama's more-or-less successful campaign for national health insurance.

Two narratives have tried to capture the welfare state's ascendancy, each strongly linked to a stage in German politics, respectively to the Second Reich and the Third Reich – where the 'First Reich' had referred to the Holy Roman Empire that spent a thousand years trying to unify Europe under Christendom. This starting point underscores the welfare state's founding universalistic aspirations, given that the spread of the Christian ideal was expressly designed to overcome ethnic and even family differences. I shall return to the significance of this point in section six. The difference between the two narratives lay in whether – to adopt Richard Dawkins' (1976) influential distinction – the welfare state is designed to have 'memes' (aka ideology) discipline 'genes' (aka biology) or vice versa. Proponents of 'Second Reich welfarism' saw ideology guiding biology, whereas proponents of 'Third Reich welfarism' would have biology steer ideology. For over a century, sociology has negotiated its *raison d'être* between these two extremes. This point can be easily seen in terms of Weber's original tripartite conception of social stratification. Much of the continuing appeal of *class* as the anchoring sociological category – even after the public discrediting of Marxism and Pierre Bourdieu's shift in the concept's economic basis from production to consumption – lies in its position midway between categories that are purely gene- and meme-driven: *status* and *party*, respectively.

The dominant narrative of the welfare state starts with Bismarck's introduction of social security insurance in the Second Reich as a political expedient to pre-empt the sort of class warfare that Marx had predicted would take place in Germany, then home to the world's largest organised labour movement. Here Bismarck, an aristocratic conservative, made common cause with bourgeois liberals in promoting the idea that all Germans, regardless of class background, shared a common fate, which had to be reflected in the terms of reference for public administration. This narrative appealed because of its 'cake and eat it' quality:It is about how a biologically heterogeneous nation-state (i.e. 'Greater Germany', which included parts of today's France, Poland, Czech Republic, Denmark, Russia) came together to support a common welfare system in which costs and benefits are borne differently by different groups, allowing each to receive according to their need and give according to their ability. As a result, both the society as a whole and all of its members are made stronger. In short, we seem to have the promised Communist revolution – but without tears.

The Second Reich narrative is often treated as the template for explaining the spread of welfare states in the 20th century. It probably works best for the United States, whose social welfare thinking is directly traceable to Woodrow Wilson's admiring account of the regime in his Princeton scholarship on constitutionalism and public administration, written a quarter-century before, as President, he introduced America to national progressive income tax, which, in a delicious piece of Hegelian irony, enabled the funding of the war that ended the Second Reich (Goldman 1952: chap. 5). A key feature of the Bismarckian animus that tends to be underplayed – but was not lost on Wilson – is that imperial expansion rested on a desire to spread ideas – rather than genes – across one's borders, especially the idea that German culture is the highest form of human self-realisation. In Wilson's hands, it resulted in a subtle but lasting shift in America's self-understanding. The nation's sense of 'manifest destiny' that for nearly three centuries had been modelled on the original English Puritan settlers' vision of America as literally the 'promised land' had metamorphosed into America as the world-historic vehicle for the promotion of freedom, democracy and 'humanity' (Wilson's own favoured word) throughout the world.

A telltale sign of the deep influence that the Second Reich has had on the American psyche is its paranoid concern with the 'ideological infiltration' of 'Un-American activities' that escalated after the First World War and, of course, continues to this day (Hofstadter 1965). In its original Bismarckian incarnation, the 'German Idea' was forever facing challenges, at first from the atavistic native threat of the Jesuits who could not accept Rome's diminished status in the secular world. But even the threat from the British Empire, which could have been defined solely in economic and military terms, was portrayed as part of a global *Kulturkampf* between properly spiritual and merely utilitarian values – albeit one in which the 'spirit' was grounded more in biology than in theology or even classical humanism (Zimmerman 2001: chap. 2). The most positive outcome of this largely self-manufactured sense of struggle was Germany's ascendancy to the premier scientific nation in Europe by the eve of the First World War. Less positively, the *Kulturkampf* set a precedent for the 'science race' between the US and USSR that became the signature battleground of the Cold War – and arguably the 'culture wars' over 'political correctness' that have now been raging within US humanities faculties for the past quarter-century.

From a biological standpoint, the Bismarckian welfare state provided an artificially sustained ecology for the co-habitation of genetically diverse peoples who have been brought together by a sense of mutual benefit in the face of common foes, be they defined as the other imperial powers, radical separatist movements within the German Reich or, increasingly, a vaguely defined but no less potent worldwide Communist menace. However, the cost of this effort was substantial for all concerned, as largely native-born citizens were burdened with subsidising the assimilation of newly absorbed peoples into expanding education and health care systems, a process that often threw up unexpected problems based on the language and history – if not the sheer physical make-up – of the assimilated peoples.

Soon after the strengths and weaknesses of the Bismarckian trajectory were revealed, a band of Darwin-inspired biomedical scientists – the 'racial hygienists' – began to argue that there are biologically prescribed limits to a viable welfare state (Proctor 1988). As Germany expanded its borders, it effectively acquired peoples with medical and educational needs that Germany had not previously faced. This imposed a greater tax burden on the native population to extend welfare coverage to the new people and their new problems. The sort of indefinite global expansion advocated by Bismarck in the name of the 'German Idea' was bound to exacerbate political and economic tensions both within Germany, between Germany and other nations. In short, Bismarck's policy was ecologically unsustainable in the long run, except under conditions of perpetual war, as imperial powers struggled over scarce resources.

However, the anti-imperialism of the racial hygienists did not stop them from supporting the idea of the welfare state. Indeed, they are the source of the second narrative of the welfare state's ascendancy, a perverted form of which came into its own during the Third Reich. Whereas Bismarck's support had come mainly from Conservatives and Liberals, the racial hygienists tended to be Social Democrats. For them, the state's sense of welfare had to be confined to biologically homogeneous peoples subject to sustainable growth rates. The policy implications were clear: no empires and no wars, with each welfare state benchmarked to a stable native population. A world government might work towards this outcome, perhaps through a global tax for the resettlement of displaced peoples and the provision of infrastructure to allow the homelands to deal with the distinctive medical and educational needs of their peoples.

This latter policy, for all its pioneer interest in what after the Second World War came to be called 'development aid', effectively let the contours of political and economic life be dictated by biogeography. (For a sense of contrast, consider the US free market economist Julian Simon, who famously argued that a more efficient means to achieve global prosperity would be for all nations to operate with an open borders policy that enabled everyone, regardless of genetic make-up, to move to wherever their skills happened to be of most use.) Accordingly, the state's 'natural' unit is the *nation*-state, which is to say, the rational organisation of the peoples who are historically tied to a region, typically as determined by parental language. It was in this spirit that claims to citizenship and territory, as well as the 'imagined communities' of nationalist myths were made throughout the 19th and 20th centuries (Anderson 1983). All of these activities were true to the original use of 'nations' to name the residence halls of the medieval universities, which collocated students who shared the same vulgar tongue. It is one of the many senses in which the university has been used as a model for governance down through the ages.

Racial hygiene received a boost after Germany's defeat in the First World War, as the interests of the native Germanic peoples appeared to have been compromised by overweening imperial ambitions fuelled by the two professions most closely associated with global expansion, financiers and scientists. As it happened, both professions included a disproportionately large number of Jews, an inveterately nomadic people with no obvious natural homeland other than the Palestinian one promised in the Bible. While today this observation tends to be regarded in the shadow of Nazism, in 1920 it was also endemic to avant-garde British intellectuals, the self-declared heralds of a 'New Age', the now familiar counter-cultural phrase that began life as the title of their flagship journal (Collini 2006: chap. 4). This biologistic strand in British social thought is the subject of the next two sections. However, I shall conclude this section with a discussion of the welfare states of Scandinavia, which resulted in a fundamental reconceptualisation of the state as such.

In the Scandinavian welfare state, the classic Hobbesian view of the state as keeper of the peace and protector of civil society metamorphosed into one of stewardship for a genetically closed population. Originally the external protector of the various families living within its borders, the state now usurped the family's traditional welfare functions as, so to speak, an 'economy of scale', effectively becoming the family writ large. This account has enjoyed perennial support amongst those who believe that the size and shape of a state is subject to biological limits corresponding to the state's effectiveness as a vehicle of genetic reproduction (e.g. Cavalli-Sforza 2000). However one ultimately judges this particular version of evolutionary social psychology, it is not unfairly seen as an *avant la lettre* extension of Richard Dawkins' (1976) 'selfish gene' hypothesis from individuals to collectives. In both cases, welfare is promoted mainly as a means to realise a larger biological end.

This general line of thought was built into Scandinavia's home grown conception of social democracy that in the Cold War era came to be seen as the 'third way' between American capitalism and Soviet Communism. The most thoughtful early version of this thinking appeared in Alva and Gunnar Myrdal's *Crisis in the Population Question* of 1934, which served to justify sophisticated domestic eugenics programmes in Sweden that *inter alia* provided economic incentives for increased procreation amongst people of normal to superior Nordic stock and compulsorily sterilised those whose performance on intelligence and motor tests revealed them to be inferior, or 'adversely selected'. Versions of these practices were in force across all of Scandinavia – in Sweden, until 1975 (Broberg and Roll-Hansen 1997). Moreover, scientific interest in 'soft racist' approaches to the biosocial divide remains strong there, as in the epigenetics research cited in the previous section, which emanates from Sweden (Hunter 2008). In this respect, Scandinavia pioneered 'national socialism' in the strict sense of a form of socialism whose sense of legitimate rule is determined by an understanding of nationhood defined along racial-scientific lines. It is the ideological formation that Hitler observed, emulated and amplified to strategic advantage.

Given, on the one hand, the culmination of Nazi policies in the Holocaust and, on the other, Scandinavia's recent relative openness to biologically heterogeneous immigrants, there remains considerable reluctance to treat 'national socialism' as an analytic category that might cover both Nazi Germany and, say, the Myrdals' contemporaneous vision of Swedish social democracy (e.g. Berman 2006; cf. Fuller 2006b: chap. 14). Here it is worth noting that the precise sense in which the Nazis 'perverted' science pertained to the forms of power that they availed to scientists in order to turn their research into policies. Generally speaking, the content of the science itself was not perverted. Rather, one would simply not expect the science to have been made such a direct basis for policy. Nevertheless, one is hard-pressed to find a reputable geneticist in the 20th century who did *not* see a role for eugenics as part of a comprehensive social policy (Kevles 1985; Pichot 2009). Where geneticists parted company was over the regimes for which they would be willing to work, which turned on their scruples concerning democratic politics and scientific fallibility.

4 Britain's curious erasure of both biology and sociology: 1907 and all that

In the previous section, I noted in passing the sympathy for racial hygiene expressed by the self-styled 'New Age' thinkers of early 20th century Britain. Generally speaking, the New Agers were either formally trained in or avid followers of the most recent natural scientific trends, which they saw as providing the basis for a distinctly 'modernist' worldview that would eventually reconstruct our understanding of humanity. To be sure, they agreed more in their sources of inspiration – Darwinian biology and Einsteinian physics – than in their policy implications. Yet, that level of concerted activity was sufficient to trigger a 'value-free science' backlash, most notably the logical positivist movement, who often appealed to the recently deceased Max Weber as their standard-bearer (Proctor 1991: chap. 10). In retrospect, this episode may be seen as the opening salvo in the ongoing 'science wars', which even in their postmodern guise remain a battle between those who would restrict and those who would extend the authority of science to non-scientific domains. An irony that might be lost on future historians is that the side typically labelled 'anti-scientific' (i.e. the 'New Agers', both then and now) is the one that would *extend* – but thereby diffuse – scientific authority (Fuller 2006a: chap. 5; Fuller 2010: chap. 4).

Amongst these New Agers were two failed candidates for the first British chair in sociology, established at the London School of Economics (LSE) in 1907: the science fiction writer and professional utopian H.G. Wells (1866–1946) and the urban and regional planner Patrick Geddes (1854–1932). Both contributors to the *New Age* had studied biology under Darwin's great champion, T.H. Huxley. They also had equally strong – yet opposing – scientifically informed views about how to deal with the 'Jewish Question'. Wells supported racial assimilation on the eugenic grounds that Jews raised the IQs of the peoples with whom they mated, whereas Geddes helped to realise the dream of a Zionist homeland as the planner for Palestine's first modern city, Tel Aviv. As it turns out, the successful candidate for the first sociology chair, L.T. Hobhouse (1864–1929), *Manchester Guardian* journalist, social liberal and Oxford idealist, did not share their New Age enthusiasms, even though they were also largely shared by the LSE's Fabian socialist founders. However, Hobhouse was the most academically respectable candidate and did not pose an intellectual threat to the statistician Karl Pearson (1857–1936), who was already ensconced in what was effectively a eugenics chair at University College London. And so the Fabians settled for Hobhouse (Dahrendorf 1995: 89–90).

I earlier observed the historical curiosity that biology had been excluded from the establishment of sociology of Britain, even though at the time Britain was the home to the most distinctive and influential biologically oriented thinkers and researchers in the world. Equally curious has been the UK's 'absent presence' in the overall history of sociology. The history of the discipline, at least understood as a body of theory, can be easily written without mentioning Britain at all – even when written by British sociologists (e.g. Giddens 1976). Symbolic of this point is that the great canoniser of modern sociology, Talcott Parsons, studied at the LSE while Hobhouse still held the chair yet felt no need to include Hobhouse as a founder of the discipline. A return to the key institutional moment – the establishment of the first sociology chair at the LSE in 1907 – allows us to relate the kind of discipline sociology has been in the past to how it might be in the future, as it engages with contemporary developments in the biological sciences. In the next section I will discuss the methodological implications of this approach for what I call a 'normative historiography of science', an updated version of Hegel's idea of history as philosophy teaching by examples, a view that goes back to my earliest work in social epistemology (Fuller 1988: Parts II and III).

One way to cut the Gordian Knot of Britain's absent presence in the history of sociology is to claim Karl Marx for Britain. It certainly has historical warrant. Marx's mature works were written in London, utilising the resources of the British Museum, often in collaboration with Friedrich Engels, whose background as heir to a Manchester textile manufacturer gave him access to the everyday life of working class people. Histories of British sociology have made much of the anchoring effect of Marx and Engels' London residency, which when combined with native reformist traditions of Utilitarianism and Low Church (typically Methodist) Protestantism laid the foundations for the discipline's progressive labour policy orientation through most of the 20th century. However, Britain's claim to Marx is a mixed blessing when projected on a world-historic stage. It reinforces the image that the nation's substantive contribution to the discipline is confined to its prehistory, with Marx deployed to show the limitations of British thought, with its supposed focus on *homo oeconomicus*, the atomised asocial individual (cf. Levine 1995: chap. 7). Thus, Britain is portrayed as the historic home of political economy, a discipline that by the end of the 19th century had divided into (neoclassical) economics, political science, geography and sociology – but only once it had been exported across the Channel to France and especially Germany. Even Parsons (1937) who for both intellectual and ideological reasons refused Marx entry into the sociological canon found a politically correct (i.e. liberal welfarist) pre-sociological political economist, Alfred Marshall, to enact the initial British moment.

It might be argued that even if it is fair to Britain to appoint Marx the nation's (proto-) sociological standard-bearer, it is not fair to Marx, given his roots in German philosophy, especially Hegelianism, and French socialism, not least the Saint-Simonian movement that also spawned Auguste Comte. But it is precisely against the backdrop of those roots that Marx appears so 'British'. As might be expected of a liberal individualist, Marx attenuated the extent to which systematic social interaction can alter fundamental cognitive dispositions. In particular, he did not treat the nation-state as the boundary condition of modern society, an assumption common to Durkheim and Weber that began to come under serious challenge within sociology only with the winding down of the Cold War and the postmodernisation of capitalism as 'globalisation'. Even the signature Marxist concept of 'class consciousness' emerges through the mutual realisation of individuals who stand similarly with respect to the means of production. The much-vaunted 'whole greater than the sum of its parts' in Marxist social theory refers mainly to the increased capacity for social transformation that results from mass awareness that one is not alone in one's plight. To be sure, that shared experience reflects economically structured social relations. But against the corporatist political logic prevalent in social democratic and fascist regimes, Marxists have typically wanted to raise class consciousness not to reinforce the existence of classes (say, through their collective self-representation in assemblies or planning boards) but to enable people to remove the structural conditions that divide consciousness along class lines (Berman 2006: chap. 9). Of course, as the history of Marxist regimes amply demonstrates, translating this intellectual subtlety into policy is easier said than done. Nevertheless, at least as a matter of principle, Marxists have shared with liberals a preference for 'voluntary associations', which may include 'mass movements', over official state action as the vehicle of social transformation.

Overall Marx has a greater spiritual affinity with his two great British contemporaries, John Stuart Mill and Herbert Spencer, than with his co-Trinitarians in the sociological canon, Durkheim and Weber.

Interestingly, Mill and Spencer are rarely – and then only grudgingly – accorded any foundational role in sociology, even though the former was among the first sympathetic expositors and promoters of Comte's sociology (which he later renamed 'ethology') and the latter brought the word 'sociology' into English usage through his own writings. (Indeed, Spencer the 'Social Darwinist' is often portrayed as a disciplinary founder of *American* sociology, due to the time he spent there in the latter part of his life promoting his books on lecture tours.) To be sure, the influence of Mill and Spencer permeated early British sociology, especially in the lines of inquiry emanating from, respectively, L.T. Hobhouse and Patrick Geddes – but typically with little formal acknowledgement of the two Victorian liberals (Collini 1979). This relative lack of acknowledgement persists in subsequent histories of British sociology. Thus, the most authoritative recent history (Halsey 2005) bundles Mill and Spencer together as literary figures who said sociologically relevant things but without engaging in proper sociological research. Is this a fair assessment of their disciplinary significance? I would say that it is one of those claims that remains true just as long as one does not draw the most obvious conclusions from it.

If my judgement seems cryptic, that is because Halsey's assessment of Mill and Spencer presupposes a certain way of reading the history of sociology – and perhaps science more generally – that treats the past more as waste product than recyclable resource. In this context, 'literary' is a predicate that sociology had to shed before it was to achieve scientific respectability, even though Halsey recognises that novelists like Charles Dickens and George Eliot were crucial in sensitising the British public to the need for systematic sociological inquiry. Mill and Spencer, of course, appear to us closer to the 'fact' than the 'fiction' pole of the literary divide. *Prima facie* they are prototypes for today's 'public intellectuals' who jostle for column inches alongside the likes of Christopher Hitchens. However, the fit is an imperfect one because today's public intellectuals live in the aftermath of academic specialisation, especially in the social sciences. Thus, academics often accuse intellectuals of misrepresenting, when not outright poaching, their own hard-earned empirical findings and theoretical insights. This has led to a two-way critical dynamic, whereby intellectuals claim to be adding policy relevance to esoteric knowledge that it would otherwise lack, while academics claim that those 'esoteric' features – often relating to matters of method – are precisely what give their knowledge legitimate authority (Fuller 2009a: chap. 3).

However, Mill and Spencer, neither of whom studied or taught at universities, encountered a rather different sort of opposition from the academic establishment of their day. Broadly speaking, they were anti-clerical: They contested the Oxbridge presumption that one had to belong to the theological mainstream to contribute in a socially responsible fashion to knowledge production (Snyder 2006). The key phrase here is 'socially responsible', a quality that both Mill and Spencer were frequently accused of lacking by their detractors. Over the previous century (say, 1750 to 1850), progress in the mechanical arts – the experimental basis of the natural sciences – had demonstrated that people from all sorts of religious backgrounds (typically non-conformist) could produce knowledge of potential use to anyone. The question was how this burgeoning knowledge base was to be organised and distributed so as to result in more good than harm for society as a whole.

Biblical concerns about knowledge in the hands of beings tainted by Original Sin clearly informed this question. But these concerns also had some obvious secular political dimensions. Tories, who generally adopted a paternalistic attitude towards society, worried that the unregulated spread of technical innovation would destabilise the social order, exacerbating already existing economic inequalities, as some people would be better positioned to benefit from the new opportunities, exploiting others in the process. In this respect, Marx's genius lay in recognising the justice of the Tory charge without renouncing the overall social value of innovation.

But his was not the only constructive response. The Cambridge natural philosopher and theologian William Whewell, who coined 'scientist' to name a specific profession, addressed the moral issue in largely cognitive terms: He proposed that competence in the mechanical arts should be academically certified by study of physics, which placed such transformative knowledge in the context of divinely inspired timeless principles of natural law. It was here that the image of Isaac Newton as someone whose piety enabled him to fathom the mind of God came to be crystallised as the archetypal scientist (Fuller 2010: chap. 3). Secular descendants of this ideal remain in what the logical positivists called the 'covering law model of explanation'. Kuhn provincialised the idea for specific disciplines as 'paradigm', which is in turn devolved into various 'peer review' processes that ensure scientists remain within certain prescribed norms of conduct (Fuller 2000b: chap. 1).

To understand Mill and Spencer's ambiguous status in the sociological canon – even in Britain – one needs to appreciate just how much they opposed this trajectory, which we now take for granted as the natural course of scientific development. In contrast, for them sociology's target audience was in government and the media – not academia. Indeed, Mill and Spencer regarded sociology as a 'science of legislation', an empirically informed successor to classical political philosophy (Collini et al. 1983). To be sure, they would make full use of the methods available to social scientists – Mill perhaps more than Spencer – but to quite different effect. Ethnographies would be commissioned, by Parliament or private foundations, not for their own sake but only on a need-to-know basis to sensitise policymakers to living conditions that require some sort of remediation or administration. Progress along such policy lines would then be charted using statistical techniques. A presumptive understanding of human nature – say, in terms of utility maximisation – would justify the categories under which the statistical data would be gathered and amongst which trends and tendencies would be sought.

This sort of sociology would not produce knowledge for its own sake, let alone knowledge that could sit comfortably with other specialist learning. Its starting points would not be underwritten by academically certified theory, and its outcomes would not be subject to review by specialist peers. Indeed, all the *a prioris* of academic knowledge production would be replaced by the *a posterioris* of the market, especially what British social science funders now call 'users and beneficiaries'. Here is a context for understanding the *modus operandi* of this alternative 'sociology'. Thirty years ago, when Pierre Bourdieu's work was only beginning to loom on the English-speaking horizon, Randall Collins (1979) proposed what he called 'credential libertarianism' as a solution to creeping cultural capitalism, whereby more and more training and vetting seems to be required to license socially relevant action. His proposal amounted to allowing anyone, however credentialed, to claim an expertise as long as the outcomes of their practice were publicised. This would allow potential clients to decide for themselves whether to take the risk and employ these self-declared experts. The balance of power would radically shift from knowledge producer to knowledge consumer, with, say, physicians judged as if they were household goods.

I believe that the fear and loathing that Mill and Spencer – and Marx – inspired in many academics of their day came from the implicit credential libertarianism of their own practice. They effectively challenged the guild-like monopoly that universities exerted over the production of legitimate knowledge. It is interesting to observe the rhetoric deployed by these theorists when debating and otherwise differentiating themselves from academics. Academics are portrayed as restricted by fixed ideas that are really social conventions promoted to the status of eternal truths. In this context, the call for openness to empirical observation variously made by our Victorian renegades was more than a narrowly epistemological proposal but an invitation to allow a broader range of human experience – and therefore experiencing humans – to count in the normative constitution of society (Kent 1981). Here Marx and Spencer appear considerably more radical than Mill, as they treated the state's representation of the social order with the same scepticism as the university's representation of the epistemic order (Offer 1999). A good current comparator to the relatively hostile response generated by this alternative 19th century grounding for sociology may be New Age science and medicine, which typically combine academic and non-academic sources of knowledge without always privileging the former over the latter, especially when the latter speak directly to the experience of a heretofore neglected constituency.

Upon turning to the contest for what became the first British sociology chair, to which L.T. Hobhouse was appointed at the London School of Economics in 1907, it is clear that amongst the three serious contenders – Hobhouse, Patrick Geddes and HG Wells – Hobhouse, who trained under the Oxford non-conformist Christian and Hegel enthusiast Thomas Hill Green, was the only one who could pass muster as a 'scientist' in the academically respectable sense canonised by Whewell (Renwick 2012: chap. 6). To be sure, Geddes and Wells had studied biology with Darwin's great champion, Thomas Henry Huxley, in what later became Imperial College London. But at the turn of the last century the referent of 'biology' was very much like that of 'sociology'. Both could refer either to a broad-based ideology (where Lamarck, who coined the word, played the role of Comte) or a loose confederation of research programmes not easily assimilated to either the humanities or the physical sciences. Indeed, in the meetings of the Sociological Society preceding the appointment, in which all three candidates participated, one issue up for grabs was whether a sharp distinction could be drawn between sociology and biology.

Looming large in the discussion was the status of 'eugenics' as something properly pursued within sociology. Although the Fabian socialists who founded the LSE looked kindly on eugenics, they were also mindful – as members of a newly federated 'University of London' – of not wanting to compete with University College London, where Karl Pearson, the *de facto* academic representative of the founder of eugenics,

Francis Galton, held a chair (Dahrendorf 1995: 89–90). Thus, the joint demands of academic respectability and institutional specialisation made Hobhouse the natural choice for the sociology chair. In contrast, both Geddes and Wells dealt explicitly with the role of heredity in society: Geddes tended towards a recognisably Germanic view that took seriously the idea of *Heimat* ('homeland'), whereby social problems could be solved by people living where they ancestrally belonged (as opposed to being forced to migrate for political and economic reasons). Nowadays we speak of 'sustainability' to remove the sense of racial memory that originally tainted this position. For his part, in terms of human evolution, Wells was less concerned with the side of the environment than that of the organism itself. He believed the future lay in cultivating what geneticist Richard Goldschmidt would later call 'hopeful monsters' (aka beneficial mutations) while culling those demonstrably defective in mind and body, amidst the wide variety of individuals thrown up by the gene pool.

It is common nowadays to treat Hobhouse as a false start to British sociology, since he remained a social philosopher throughout a career whose influence did not seem to extend beyond his immediate successor Morris Ginsberg. However, at least when compared with his two rivals, several features of Hobhouse's vision seem to have had lasting impact on British sociology (Renwick 2012). The first is the wedge he drove between biology and sociology, which not only codified the division of labour at the University of London but also corresponded to developments in institutionalising sociology in France and Germany, albeit occurring under rather different political circumstances. Basically, biology dealt with the prehistoric and sociology with the historic – but not descriptively, as a historian would, but prescriptively, with an eye to the future. Thus, Hobhouse took it to be a 'sociological' insight that the state had to complement the market by regularly redistributing wealth so as to keep capital fluid and hence the economy dynamic. 'Sociology' here meant a systemic view of social life with a clear sense of what counts as optimal and suboptimal functioning, the difference between which might be remedied by empirical enquiry resulting in policy proposals.

In this respect, Hobhouse, whose pre-LSE career was spent at the *Manchester Guardian*, might be seen as bridging the sociological worldviews of Mill and Parsons. This spirit, including the aversion to biology, continues in the work of Anthony Giddens, despite having turned his back on sociology's Anglo-American roots (especially Parsons) in favour of continental Europe. Like Hobhouse, Giddens' reputation emanates from the classroom and the newsroom rather than the research site. Moreover, a core constituency for both has been the social work community, whose significance remains underplayed in histories of British sociology – meriting only one page in Halsey (2005). In Hobhouse's sociology, social workers enabled the state to compensate for the deficiencies of the market through interpersonal mediation with those most in need. (In Giddens' jargon, this would count as forging the structure-agency nexus.) In effect, for Hobhouse social workers functioned as secondary school teachers did for Durkheim, namely, a cynosure in the overall project of maintaining stability in a rapidly changing society, what they – and Giddens – called 'citizenship', which, with some justice, may be brutally assimilated to 'stakeholdership'.

In contrast, Hobhouse's two rivals for the LSE chair did not seriously include the spontaneous meaning-making capacities of humans within their conception of sociology, Thus, Geddes' 'social surveys' regarded people largely from a behavioural standpoint, as an ecologist might observe an animal population in its native habitat, while Wells, whose worldview I shall explore below, targeted his version of sociology exclusively to elite policymakers. Neither was really interested in engaging with the opinions of the people whose lives would be purportedly improved by the appliance of sociology. Such antipathy to ordinary subjectivity probably related to their scepticism about the reliability of any data generated in such an empirically restricted and self-serving fashion. That people might hold certain beliefs about their social world did not carry *prima facie* evidential value for either Geddes or Wells. Rather, the authority claimed for such folk beliefs reflected a certain moral standpoint, perhaps influenced by lingering religious conceptions of the soul. In any case, these should not needlessly trouble a sociologist interested in designing an infrastructure that might realistically ground a 'modern' sense of social order. In the end, Geddes and Wells believed that social science was simply the final frontier of 'natural science', however differently they understood that phrase.

In recent years nostalgia has developed around Patrick Geddes as British sociology's lost hero (Fuller 2007b: chap. 5). Whatever the intrinsic merits of this proposal, the move is symptomatic of our own renewed willingness to renegotiate the disciplinary boundary between sociology and biology. Behind this are genuine scientific developments over the past century, especially a finer grained understanding of how both genetic and environmental factors influence the human organism throughout its lifetime, rendering any textbook distinction between 'nature' and 'nurture' increasingly difficult to defend. But equally it reflects the somewhat euphemistic resurgence of the constellation of 'blood and soil' ideas associated with geographical determinism – in which 'race' had figured as a scientific principle – only now in terms of such soft-focus categories as 'genetic diversity', 'carrying capacity' and 'sustainability'.

Interestingly, in their original incarnation, these ideas fed into the first great period of climate change theorising, a century ago at the peak of imperialism, in which Geddes amongst others argued for smaller self-organising social units, repatriation and controlled population growth as solutions (Stehr and von Storch 2000). Finally, the financial incentives for sociology to hitch its fate to that of the lab-based biomedical sciences cannot be overlooked as inviting a second and third look at overlapping cross-disciplinary research concerns. To be sure, these are not the fields of biology that especially interested Geddes, whose own naturalistic sensibility resembled that of Darwin himself. But they were very much at the heart of the biosocial concerns of H.G. Wells, to whom I shall now turn.

Wells is an awkward figure to discuss in the context of the history of British sociology, and not simply because of his reputation as a novelist and polemicist who wore as a badge of honour his lack of factual sociological knowledge. More specifically, in the period leading up to Hobhouse's appointment of the first sociology chair, Wells regularly lambasted the Fabians on 'the so-called science of sociology', to recall the title of his widely noted 1906 address. By this he meant to deride those who rallied around the banner of 'sociology' in the Anglophone world at the time, namely, followers of Comte and Spencer. It was precisely against this sense of 'sociology' that Wells counterposed his own science of 'utopia' (Lepenies 1988: 145–54).

Commentators have found it difficult to discern Wells' substantive intellectual point beyond ridiculing the lack of imagination and vision he found in the empiricism practiced by self-avowed sociologists. Perhaps such commentators do not take seriously that even 100 years ago the sense in which sociology should be 'empirical' was still an open question. Given Wells' defence of utopia, a better way to see what he opposed is in terms of what Karl Popper (1957) would later call 'inductivism' and 'historicism': the doctrine that, on the one hand, the future repeats the past; on the other, the past necessitates the future. It is easy to see how a pursuit of social facts for its own sake – rather than for simply establishing a basis to criticise taken-for-granted attitudes and policies – would make one susceptible to these Popperian sins. If Comte and Spencer prove villains for Wells, it is less for their own mindsets – after all, neither could be accused of lacking imagination or sticking too close to the facts – than the air of finality to their views that encouraged followers simply to reinforce, perhaps with minor amendments, their schemes with relevant factual details. In short, Wells was revolting against the cognitive propensities that Kuhn identified with 'normal science'.

Wells traced his own science of utopia to Plato, Thomas More and Francis Bacon – as well as those aspects of Comte's own thought that drew upon them. But how exactly to characterise the intellectual lineage defined by these thinkers? To simply call it 'literary', as Lepenies (1988) and Halsey (2005) are inclined to do, is to conjure up much too readily an aversion to the scientific method. If anything, these thinkers, not least Wells, were more open to the charge of overextending the scientific method, aka 'scientism'. It would be closer to the mark to say that utopian thinkers relate to facts differently from more pedestrian practitioners of the scientific method, the people Wells pejoratively called 'sociologists'. In terms of cognitive functions, the utopians see facts as appealing more to the imagination than the senses. For them, facts do not constitute truth as pieces in a puzzle but are rather the raw material from which truth may be 'distilled' in a sense to be explained below.

Nowadays this attitude is most closely associated with Roy Bhaskar's 'critical realism'. It presupposes that our ordinary ways of understanding the world, summed up as 'empiricism', produce an epistemic alloy of reasonably reliable observations skewed by preconceived ideas, the result of which encourages us to do little more than adapt to current expectations. At this point, the critical realist and the utopian share a common strategy: They proceed *not* to remove all bias from these observations but to replace the backward-looking ideas informing them with more forward-looking ideas – ones that Bhaskar (1986) would call 'emancipatory'.

This replacement strategy requires that the truth be 'distilled', a process most vividly illustrated by the controlled experiment, the paradigm case of the scientific method, whereby one tries to demonstrate a particular factor's contribution to a reliably produced outcome. The utopian is especially interested in manipulating the influential factors so as to enhance the 'good' and diminish the 'bad'. Consider a population of individuals deemed 'successful' along a given dimension. Some may share a common quality despite the other differences in their lives, whereas the success of the rest appears to have resulted from a concatenation of different factors in each case. The utopian would dwell on the former group, while dismissing the latter as lucky.

Wells, who thought in precisely these terms, developed a lifelong interest in eugenics, the centerpiece of sociology as a utopian science. At the same time, his views about which genetic theory should inform eugenics changed as the scientific frontier shifted, moving from Galton to Mendel in the early 20th century (Bowler 1989). I stress this point because the disastrous medical and agricultural policies associated with Nazism and Lysenkoism have resulted in a widespread stereotyping of eugenics as misapplied science in a sense that conflates the epistemic and ethical meanings of 'misapplied'. In the main, eugenicists were just as sensitive to cutting-edge genetics research as ordinary policy-makers might be to the latest work in sociology or economics. The eugenicists' fault – assuming that it is a fault – lay not in a doctrinaire attitude towards a particular genetic theory but in the very idea that genetic theories should influence policy interventions.

Of the three contenders for the first LSE chair, two of them – Geddes and Wells – included eugenics within the purview of what they called 'sociology'. Of these two, Wells' version of eugenics more clearly anticipated today's 'biotechnology', as the name of a commercial industry that typically prioritises the sheer production of novel organisms over the provision of adequate sustaining environments for them. In contrast, Geddes conceptualised eugenics in the spirit of a gardener who aims for managed variety in nature by bringing into existence only those novelties that are likely to be sustainable with the already extant species. Nevertheless, Geddes seems to have introduced 'biotechnology' into English (from the German *Biotechnik*), but for him it clearly meant something like 'ergonomics' (Bud 1993: 69–70).

As for the successful candidate, Hobhouse acknowledged the relevance of genetic variation to the social order without embracing a proactive science of eugenics. It was here that Hobhouse remained more liberal than socialist. He followed Adam Smith who rationalised human diversity through a spontaneous division of labour in which everyone comes to value the different talents of their fellows by consuming their products and services. In this process, the state's role is not to reduce or bias such diversity, as a eugenicist might, but to remove economic barriers to its full expression in the marketplace.

William Beveridge (1879–1963), the economist who became director of the LSE in the latter half of Hobhouse's career (1919), went on to design the British welfare state on elements of Hobhouse's and Wells' vision that were grounded in the imaginary foundational social science he called 'social biology'. This fantasy discipline, while ultimately a failure in its own right, became the springboard for modern conceptions of 'equal opportunity' and 'meritocracy' as regulative principles of postwar social policy (Renwick 2011). Social biology accepted the political premise of Hobhouse's 'social liberalism', namely, that the state is designed to enable humans to be all they can be. This justified the redistribution of both earned and unearned wealth to provide an adequate level of health, education and subsistence. But equally, social biology was mindful of Wells' view that even once everyone is provided this basic level of social security, it does not follow that each will be able to contribute equally well to society. For Wells, eugenics would take over at this point, eventually converting Plato's 'myth of the metals' in the *Republic* – the organisation of society by the stratified cultivation of talents – into a scientifically determined aristocracy.

It is easy to imagine a pure version of this vision: The welfare state over time would result in less, not more, cross-generational social mobility, as people's natural capacities are with increasing accuracy identified, tracked and matched to appropriate forms of employment. To be sure, an ineliminable element of social mobility would always remain, as the statistical nature of the laws of heredity occasionally throws up mutations whose fates cannot be foreseen but whose prospects for reorienting the entire social system might be great. It was this openness to such 'hopeful monsters' – the eugenic equivalent of the world-historic hero – that prevented the Beveridge-Wells meritocracy from being merely a scientific version of the caste system. Nevertheless, it also made at least Wells suspicious of the self-protective tendencies of parliamentary democracy and susceptible to the charms of the self-affirming dictator – qualities shared by the more intellectual defenders of National Socialism, such as the jurist Carl Schmitt (Coupland 2000).

The view of the social order that made bedfellows of social democrats and fascists in this fashion might well be called 'punctuated equilibrium', to recall the phrase coined by Stephen Jay Gould in the 1970s for his alternative to Darwin's gradualist theory of evolution. However, there was a crucial difference between the likes of Wells and Schmitt: The Nazis conceptualised the need for dictatorship – the 'punctuation' – to come from external threat, whereas the social democrats saw it as internally driven by the probabilistic character of genetic transmission.

However, the person appointed to the chair in social biology in 1930, the zoologist and science populariser Lancelot Hogben (1895–1975), was set on distancing progressive – in his case, specifically Marxist – politics from eugenics (Dahrendorf 1995: 249–66). He had the advantage of being much closer to the research frontier of genetics than either Beveridge or Wells. This led him to conclude that the field's successes were limited to easily controlled populations in the laboratory and the agricultural station. Moreover, Hogben had just moved to the LSE from the University of Cape Town, where he actively opposed the nascent Apartheid ideology then championed by the scientifically minded politician, Jan Christiaan Smuts. Smuts coined 'holism' to capture in geographical terms the sense of functional differentiation associated with organic development, which led him to defend the segregation of the races according to habitats as an emergent feature of evolution (Bud 1993: 65–6). Smuts saw Apartheid as a version of biological 'mutualism', the process whereby two species benefit in their distinctive ways from joint activities.

It is worth observing that Geddes sympathised with this approach under the rubric of 'racial hygiene', which guided his efforts at town and regional planning in India, Palestine, as well as South Africa: that is, to create spaces that enabled different races to interact productively while retaining their territorial integrity (Meller 1994: 277). For his part, Hogben dismissed Geddes's fetishisation of a 'sense of place', arguing that social progress generally entailed the free mobility of people who ultimately bend the physical environment to their will (Bud 1993: 76). Whereas the inscrutably apolitical Geddes opposed both capitalism and socialism as ecologically unsustainable, Hogben held fast to the Marxist line that capitalism is a necessary precondition of socialism. Given today's environmental preoccupations, it is easy to see why Geddes would be riper for revival than Hogben.

All told, however, Hogben's opposition to eugenics targeted mainly its original Galtonian assumption – reiterated by Beveridge and Wells in their early Sociological Society contributions – that the biggest genetic threat to modern society came from the incompetent out-reproducing the competent (Bellamy 1992: 55–6). Hogben saw matters quite the other way round: The Janus-faced nature of scientific progress in the 20th century meant that, on the one hand, improved medical care had resulted in an overall decline in birthrates, while on the other, improved military technology had enabled increasingly efficient forms of mass murder. Taken together, the two trends would eventually depopulate the planet. Thus, as opposed to the restrictive policies normally associated with eugenics, Hogben promoted an across-the-board commitment to what he called 'biological invention' that ranged from improving fertility and decreasing mortality in humans to breeding new plant and animal hybrids for agricultural purposes (Bud 1993: 74–5). Hogben's legacy to sociology was the establishment of a non-Galtonian school of demography, which was institutionalised at the LSE by his student David Glass, the force behind the tradition of social mobility studies that epitomised postwar UK empirical sociology.

Hogben's brand of demography aimed to recover the discipline's roots in 'political arithmetic', a phrase coined in the early meetings of the Royal Society – that is to say, before economics and biology had become separate disciplines. Running through the introduction to Hogben (1938) is the sense that it had been a mistake to divide the meaning of 'inheritance' into a specifically biological and economic component, especially taking together developments in Mendelian population genetics and Marxist political economy. Hogben regarded Mendel and Marx as having scientifically corrected the unnecessarily gloomy speculations of Darwin and Malthus, respectively, which at the time were informing racial hygiene policies (28–31). Specifically, Mendel corrected the Darwinian tendency to suppose that an individual human contains the potential to be no more than he or she already is, while Marx redressed the Malthusian assumption that humanity is at a collective loss to change substantially the physical parameters within which it survives and thrives. In short, contrary to the classic eugenicist aims of contained population growth and a stable social order, Hogben treated growth and mobility as mutually reinforcing virtues that would eventually enable society to break free of its Malthusian-Darwinian shackles.

5 Methodological interlude: Towards a normative historiography of science

The story ends with Hogben, who vacated the chair in social biology in 1937 once the Rockefeller Foundation refused to continue funding for his research, which occurred against a backdrop of antipathy from the LSE social science community – and not only from Beveridge, who resigned the school's directorship at the same time, and fellow socialists like the political theorist Harold Laski (Dahrendorf 1995: 263–6). The chair was never filled again. Dahrendorf attributes the intellectual side of Hogben's failure to his bloody-minded empiricism, which presumed disciplinary differences could be traversed by enough clearly presented facts. I believe this is much too shallow as a diagnosis of someone with such clear Marxist theoretical commitments and sophisticated grasp of the history of science. A better account is that, despite his lack of enthusiasm for eugenics, Hogben's general intellectual orientation and career trajectory placed him much closer to Wells as a throwback to the 'science of legislation' approach to sociology exemplified by Marx, Mill and Spencer.

This helps to explain Hogben's (1938) interesting rhetorical strategy of legitimising his version of demography by leapfrogging back to the

17th century, when 'political arithmetic' was advanced as a discipline innocent of later distinctions between the natural and social sciences, and then writing as if those distinctions never got made yet the subsequent history of science took place pretty much as it did. Thus, Malthus, Darwin, Marx and Mendel remain as significant figures but their achievements look somewhat different. This new look, in turn, provides the lens through which the vital statistics gathered by Hogben and his co-workers were to be seen. Rather than being atheoretical, Hogben was operating with a legislator's sense of 'theory', which is not specific to an individual inquirer or even a discipline – as Dahrendorf and other social scientists today might suppose – but to a general vision of the direction in which society should be heading. In that case, the facts properly assembled suggest underlying tendencies that may be enhanced or diminished, depending on the policy's likely bearing on the desired direction.

I have identified this attitude towards theory, shared by Hogben and Wells, as 'critical realist'. I do this advisedly in light of the stray associations that have accumulated around the position in its various stereotypings in the social theory literature. The crucial point for my purposes is that the critical realist identifies the experimental method – with its promise of showing the difference that a factor makes to an outcome – with human agency as such. The capacity to demonstrate how things might have been and still might be is integral to who we are, namely, world-creators: in Augustinian terms, *imago dei*; in Marxist terms, *homo faber*. In this respect, critical realism relates to social constructivism more as metaphysical presupposition than epistemological opponent. Our social constructions only work because we ourselves are constructed to make sense of how the world works. Although critical realists may be loath to admit it, their patron saint is Francis Bacon, in whose spirit Mill's canons of induction were proposed.

This point may not be obvious because 'we', including the founding critical realist philosopher Roy Bhaskar (1975, 1986), associate the 'experimental method' too closely with the logic of laboratory practice. Yet Bacon lived in the early 17th century, perhaps as much as two centuries before laboratory-based experiments were generally accepted as reliable vehicles of knowledge production. Indeed, his own inspiration, Galileo, was suspected of having falsified his experiments. Moreover, some of these suspicions seem to have been borne out, at least insofar as Galileo failed to distinguish between real and 'thought' experiments (Feyerabend 1975). Yet, thought experiments need not be unreliable as 'anticipations of nature', to use the old Kantian phrase, if they are conducted in a sufficiently rigorous fashion. Indeed, they have been productive throughout the history of science, not least when Einstein discovered relativity theory after imagining himself travelling on a light beam.

This observation may be extended to all counterfactual arguments, whereby at stake is how the world would be, were one or more causal factors systematically altered (Fuller 2008b). While Bacon certainly promoted the construction of laboratories as a nation-building project (the proverbial 'House of Solomon'), he also believed that the experimental method could be already applied to evaluate and inspire the disparate knowledge claims on offer from astrologers, alchemists, physicians and, more generally, natural philosophers. In practice, he meant that verbal arguments that included reports of observations could be evaluated by people not positioned to observe the events in question for themselves. A natural descendant of Bacon's broad but disciplined sense of experimentation is simulation, whereby methodical re-enactment is outsourced from the evaluator's mind to a computer programme.

Histories of the experimental method tend to presume that the conduct of science in laboratories is its natural realisation. From a rhetorical standpoint, that may well be true: The outcome of a lab-based experiment confers a sense of concreteness that bolsters confidence in its relevance to matters outside the lab. However, as the French physicist Pierre Duhem (1954) famously observed, the laboratory does no more than realise an imaginary abstract relation in a concrete setting. For Duhem, this was an argument for not necessarily discounting a theory simply because it failed to be borne out in a lab-based experiment. After all, we may have failed to do justice to the theory by not constructing or interpreting the experiment properly. For this reason, matters of 'reliability' and 'validity' are clearly distinguished in research design: that is, the difference between whether something works on its own terms and in some larger setting. Duhem's observation applies equally to other ways in which the experimental method might be realised – not only thought experiments and computer simulations but also science fiction, all of which presume the potential efficacy of counterfactual reasoning (Fuller 2010: chap. 2).

I mention science fiction because Wells clearly saw this emerging genre as the natural continuation of utopian political philosophy, to which he would have dedicated the science of sociology, had he been appointed to the LSE chair in 1907. Moreover, contrary to its portrayal in histories of sociology, Wells' perspective was not idiosyncratic. Much popular science writing today – especially related to sociobiology and evolutionary psychology – displaces academic sociology in ways Wells would have welcomed. The missing link between then and now is the so-called visible college of socialist natural scientists, typically lab-oriented biologists, whose socialism – be it liberal or Communist – entailed a scientisation of the human condition (Werskey 1978). They were amongst the leading British public intellectuals in the middle third of the 20th century, and they acknowledged Wells as a forebear, some of them even admiring his authoritarian approach to politics (Fuller 2007c). It was also the crowd in which Hogben travelled.

The 'visible college' included such figures as J.D. Bernal, J.B.S. Haldane, Joseph Needham and Julian Huxley, all of whom to varying degrees turned their back on the increasing professionalisation of science, despite having made significant scientific contributions. Like their 19th century precursors Comte, Marx, Mill and Spencer, they shared a keen interest in the history of science, mainly in the spirit of distinguishing progressive from retardant factors on the path to some emancipated future. Indeed, Bernal is arguably credited with having founded the history of science as a discipline specifically concerned with the interplay of social and intellectual factors (Fuller 2000b: chap. 7). Bernal and Haldane are especially interesting in the context of my argument because they explicitly followed Wells' example in the 1920s and wrote science fiction novels of a hypothesised future, which in Haldane's case elicited a critical but equally science-fictional response from Bertrand Russell (Paul 2005).

Setting aside the prescience – or not – of these works when it comes to genetic transformation and more radical future embodiments for humanity, they provide the trace of what remained of sociology's original non-academic impulse after much, if not most, of it had been co-opted by Hobhouse's LSE appointment. Fictional works by Aldous Huxley, C.P. Snow, Arthur Koestler and, of course, George Orwell may be seen having continued this 'subaltern' tradition. The interesting question for us is why these novelists are not normally considered writers of *sociology*?

The answer, I would suggest, has little to do with their failure to rely exclusively on strict fact. After all, most 'social theory' today is just as aromatically related to empirical phenomena as most science fiction. Rather, the difference lies in the lack of appropriate accountability for science fiction works. There is little incentive for science fiction writers to critique, let alone re-do, the visions of the future they draw from their counterfactual appraisals of history. Each writer tends to strike out on his or her own. Consequently, the enterprise has no collective direction, and it is difficult to decide the relative merits of works, because the critics who do propose evaluative criteria are loosely coupled to the enterprise. For Kuhn (1977), this marks science fiction as more an 'art' than a 'science'. But note that all of these comments about the exclusion of science fiction from sociology pertain less to its content than its institutionalisation. In other words, *à la* Karl Popper, if we were to treat science-fictional propositions as revisable hypotheses rather than stand-alone fantasy worlds, then they could quite quickly form a kind of sociology – which is perhaps what H.G. Wells had hoped would happen. The difference between the two prospects boils down to whether how one fills in what the author leaves unspecified: Does one simply imagine that it is already the case or take into account what it would cost to make it the case?

In any case, history can be used a resource for opening up future horizons for the scientific imagination. In the previous two sections this has been done by returning to a moment in the past when a decision was taken between options that could have reasonably led history in a radically alternative direction. Obvious candidate moments are ones that in retrospect can be seen to have anchored a process of institutionalisation, as in the appointment to the first chair in sociology at the LSE in 1907. However, such 'turning points' can pose special challenges to counterfactual historiography when the institutions founded in their wake remain legitimate. Although Hobhouse appears very old-fashioned today, his two main rivals for the chair, Geddes and Wells, still appear quite implausible as serious alternatives. But that is only because the discipline they would have established under the rubric of 'sociology' would not resemble today's discipline. Nevertheless, while it would take considerable imagination to piece together how these alternative sociologies would have developed, the effort would not be wasted as sociology once again renegotiates its relationship to biology, not least in a 'transhumanist' direction, which will be raised in Chapters 2 and 3.

6 Memes *vs*. genes: Humanity's perennial need to decouple ideology from biology

However justified is the charge that Richard Dawkins (1976) has simply re-invented social contagion theory with his coinage of 'memes', these germ-like ideational entities recall a sensibility that in the 20th century led both imperialists and revolutionary socialists to believe that concerted ideological infiltration could conquer biologically reproduced distinctions and prejudices, as represented by national borders and class markers (cf. Fuller 2009a: 99–106). Put in broadest metaphysical terms, this was the ultimate campaign for mind to conquer matter. The stress in Rudyard Kipling's imperialist slogan should thus be placed on the *burden* of 'white man's burden'. After all, as the liberal economist John Hobson had observed in his 1902 landmark study, *Imperialism*, if Britain, France or Germany were mainly interested in improving the welfare of their native populations, they would never have engaged in such extravagant overseas adventures. Clearly something more strictly ideological was at stake in imperialism.

Lenin (1948) famously construed imperialism as simply underscoring the 'ism' in 'capitalism'. It meant that capital was alienated from the production and consumption of goods to an endlessly expansive source of elite financial speculation – that is, capital as an idea pursued for its own sake. But sensitive to the 'cunning of reason' that Marx had inherited from Hegel's idealist philosophy of history, Lenin was equally aware – as Comte had been *vis-à-vis* Roman Catholicism's positivist potential – that imperialism's globe-spanning social structures, abetted by steamships, railroads, telegraphy and later telephony, could be turned to the advantage of the Communist revolution, even serve as its launch pad. However, the revolutionary cells, instead of existing symbiotically in the sort of formally recognised 'spheres of influence' reserved for traders and diplomats, would be housed covertly in, say, universities, living parasitically off their imperial hosts.

The idea that memes and genes might constitute competing flows for reproducing 'the human' is ultimately traceable to the religious practice of *proselytism*, whereby Christians and Muslims have engaged in worldwide campaigns to convert not only the uncommitted but also adherents of other faiths, typically those in which the family is the default source of religious affiliation. In this latter context, especially strong moral force is attached to the idea that one expressly *decides* to believe in a particular deity – that is, not simply to allow one's allegiances drift along in the direction of one's genes, as suggested by the phrase 'cultural Jew', which implies someone who was born to a Jewish family and retains Jewish customs insofar as they do not interfere with a 'normal' secular existence, given that the specific relationship with God has been severed. Such a view is abhorrent to the proselytiser because it blends genetic and memetic transmission, such that parentage is made the default source of one's ideational identity, effectively taking matters of divine commitment out of the offspring's own hands. The numerous controversies in the history of Christianity related to the infant baptism bear directly on this point.

As the above discussion of proselytism shows, the disentanglement of genetic and memetic flows in the constitution of the human is normally discussed as simply a matter of releasing the memetic from the genetic – in this case, by decoupling religious from family commitment. In those cases, the genetic yields to the memetic as the individual is conceptualised as a relatively abstract, self-creating entity – if not quite a 'blank slate' – who decides which sort of person they shall try to be and to whom they shall be accountable along the way. How-ever, memetic-genetic disaggregation might also be expressed as the prohibition against mingled memes resulting in mingled genes. This attitude was much more explicit in French than British thought of the late imperial period – namely, that the superiority of European ideas should not be taken as an invitation for colonised peoples to contaminate the European gene pool through miscegenation. The fundamental hypocrisy of this position was revealed by Frantz Fanon at the dawn of the postcolonial period – a sort of 'racist assimilationism', the internalisation of which Pierre Bourdieu recognised in his early field work under the guise of *déracinement*, whereby the French imperialists literally 'uprooted' the native North African culture by degrading it in various ways, only to replace it with the French culture, which the natives came to admire – but only at a permissible distance that respected racial differences. Indicative of this phenomenon was the need for Fanon himself to receive the endorsement of Sartre in Paris before being embraced by the intellectuals of Algiers (Grenfell 2005: chap. 2).

The historically vexed question of the social standing of women has probably done the most to call the coupling of memetic and genetic flows into question. The discussion here returns us to the religious roots of social life. In his only sustained discussion of women in *The Sociology of Religion*, Max Weber observes their paradoxical position in the transmission of religious identity (Weber 1963: 104–5). On the one hand, women are typically more open than men to the prophet's call; yet on the other hand, they are also mainly responsible for the reproduction of religious rituals. Taken together, the paradox suggests the provocative hypothesis that women in their capacity as, so to speak, the household's 'head of government' – even if not its 'head of state' – obstruct the opportunity for free religious choice amongst males. This would help to explain the frequent hostility to women found amongst Muslim and Christian proselytisers, notably St Paul (Weber's own example), which is not consistently matched by the more nuanced treatment of women in their Scriptures. But perhaps equally it explains the subtle appeal of the otherwise esoteric writings of Judith Butler (1990), whose queer theory provides a strategy for systematically divesting gender identity of the residual biologism of sexual identity.

Living in a time when oppression on the basis of sex and race – or gender and ethnicity – tends to be seen as the twin products of white male supremacy, it is disorientating to consider that someone like St Paul might have had a principled objection to women exercising power over the household because of its potentially racist implications for the transmission of ideas. Yet, at the very least, women were well placed to provide a conservative bias to the reception of new ideas by noting how they would likely disrupt the household. Feminist sociologists and anthropologists following Nancy Jay (1992) have used this point to explain the cross-cultural appeal to radical acts of male sacrifice as propaedeutic to religious renewal. Here I would observe that the need for such violent practices concedes women's religious authority under normal circumstances. In this respect, the Christian and Muslim proselytisers wanted to make it attractive to turn away from one's family to acquire a different identity, one purportedly more profound and of universal import. It should come as no surprise, then, that the woman in whom Jesus took most active interest during his ministry was Mary Magdalene, someone variously cast as a prostitute, an artisan and an epileptic – but certainly not a wife, mother or any of the other traditional female agents of societal reproduction.

Much more could be said about the gender division of labour but the following comments will have to suffice. From today's standpoint, women are most severely disadvantaged in terms of their place in politics and the economy outside the home. By contrast, in the ancient world, the primary locus of concern was the household, the security of which then allowed for the more extravagant political and economic ventures in which men dominated. It is easy to overestimate these exceptional activities today simply because they are the ones that happen to have left a strong paper trail, and *we* tie matters of significance to a written record more than ever before. But there is no good epistemological reason to think that, especially when it comes to understanding the past, evidence is proportional to significance (Fuller and Collier 2004: chap. 6). The relative lack of a record of women's exercise of power may well have had to do with the routine rather than episodic nature of that power. The exceptional tends to be recorded, while the normal is literally 'uneventful' – hence, the 'inscrutability of silence' that besets the social epistemologist interested in reconstructing a lifeworld (Fuller 1988: chap. 6). Aristophanes drove home the unspoken but real power exerted by women in ancient Athens to great comic effect in *Lysistrata*, whose plot turns on women withholding sex from men in order to end a senseless war.

However, the gender balance of power came to be decisively disrupted by the Industrial Revolution, as machinery gradually mediated, if not outright replaced, jobs done at home, Thus, the locus of women's power dissipated as food and clothing was purchased instead of made and capitalist imperatives drove the servants whom women had managed from domestic to industrial labour. Moreover, as the locus of production shifted from the farm to the factory, the home gradually lost its self-sufficiency as an economic unit and came to be absorbed as an element of 'the market', itself a subsystem of a 'society' governed by a nation-state in which men clearly ruled. This was the context that spawned the discipline of 'political economy', competence in which was biased towards the negotiating skills of males in the market place, where – in the Ricardian phrase – one angled for 'comparative advantage'.

At that point, roughly corresponding to the time of Mary Wollstonecraft's *Vindication of the Rights of Women* (1792), the freedom and power of women came to be benchmarked to what at least some men already enjoyed in public settings, and the physical and emotional labour traditionally performed by women in private life came to be devalued even by women. Wollstonecraft was herself especially artful in casting the issue. In seeming agreement with Rousseau's call in *Émile* for women and men to receive the same education so that men will no longer be spellbound by women, Wollstonecraft observed that indeed women would fare better by learning to direct their own lives rather than the lives of men (Colley 1992: 273–4). Increasingly human biological reproduction came to be treated on the model of agriculture in classical political economy, which is to say, not the exclusive dominion of its hereditary female caretakers but a sector of society in need of rationalisation for the public good. In practice this gave male politicians and economists the prerogative to introduce disciplinary standards – associated with the nutrition, education and behaviour of both parents and children – that effectively 'de-feminised' childrearing. To underscore the continuities with agriculture, an academic field called *puériculture* emerged with the first wave of eugenics in France in the 1860s and spread to several European countries, significantly due to Clémence Royer, a feminist campaigner who was also Darwin's French translator (Hecht 2003). Royer regarded the de-feminisation of childrearing as one of the most important developments in the release of women's potential from its biological captivity. But arguably all that it did was to allow 'scientific' men to colonise what had been previously a protected market for women's social expertise.

Based simply on the entangled histories of eugenics, the welfare state and the women's movement, it might be easy to conclude that a biologically oriented social science is bound to be anti-humanist, if not downright inhumane. However, this impression would be mistaken.

At most one may conclude that biology tends to *relativise* our sense of what it is to be human – and, more to the point, who counts as 'properly' human. Even though animal welfare defenders such as Peter Singer are nowadays inclined to take the overwhelming genetic overlap between our own and other animal species as naturalistic grounds for species egalitarianism, historically such a reductionist perspective has favoured an anti-essentialist view of species altogether, which effectively transfers human distinctiveness from the genetic composition of our bodies to our general capacity to compose bodies out of genes. In this respect, the ease with which eugenicists have been willing to do violence to both human and animal bodies contrasts sharply with the conspicuously non-violent approaches of animal rights defenders. The former suggests a godlike confidence that is captured by the phrase 'second creation', which appeared in the title of the first authorised account of the ill-fated life of Dolly, the first cloned sheep (Wilmut et al. 2000).

The confidence – if not optimism – of eugenics supporters is worth underscoring, even as they frequently acknowledged technical limitations and morally chequered consequences. Again this contrasts with the caution – if not pessimism – of most defenders of animal welfare (also including Singer, in other moods) whose gaze is normally shifted from the gene to the individual organism: from what can be seen only with optical enhancement to what can be seen with normal vision. In that case, the human-animal link is forged in terms of manifest sentimental attachment rather than hidden common constitution. Here humans inhabit the same ontological plane as animals, subject to the same set of vicissitudes, which in turn help to foster a sense of mutual dependency. On this view, there is no escape for humans to a higher-order plane of 'biotechnologist'. What is required then is an ethic fit not for all-powerful creators but for vulnerable creatures (MacIntyre 1999). The genealogy of this quite literally down-to-earth approach to humans might start with Donna Haraway's *Companion Species Manifesto* (2003) and reach back to the ancient Greek school of Cynics, all concerned with recalibrating our sense of humanity in terms of canine modes of being. Most recent sociological attempts to establish a positive presence for non-human beings in the social world have operated with a similarly levelled normative horizon (Franklin 1999).

But such levelling points to a curious paradox *vis-à-vis* what might be called the 'deep history of biology'. It trades on treating the relevant non-human actors as closer to *type specimens* than *population members* (Hull 1989). Thus, actor-network theory makes much of the 'heterogeneity' of non-human actors, which rhetorically privileges differences between beings in *kind* over those of *degree*, as one might expect of social researchers who follow whole actors in the field rather than observe their parts under a microscope (Bijker and Law 1993). These differences in kind are then used to account for the non-human actors' opacity, elusiveness, if not outright resistance to human attempts to assimilate them to a common social regime, resulting in the need for a 'politics of nature' (Latour 2004). And while such a move is now quite familiar (especially in 'deep ecology' circles), it effectively reverses the direction of the history of biology, where the type-population distinction normally marks the transition from 'essentialist' (aka creationist) to evolutionary thinking about the concept of species (Mayr 1970).

The type orientation imagines species as eternally fixed, with each individual a more-or-less reliable specimen of some ideal type. In contrast, the population orientation imagines species as individuals whose various differences are superseded by common features that can be reproduced to enable their collective survival. Whereas species-as-types tend to be defined *a priori* in terms of conceptually defined properties, species-as-populations tend to be defined *a posteriori* in terms of historical and geographic cohabitation. In effect, types *are* species, but populations *do* species. To his credit, Latour (1993) has conceded the pre-modern roots of actor-network theory's ontological levelling in its refusal to acknowledge a common standard for describing and evaluating the various beings. Yet, such metaphysical relativism fails to acknowledge modern political economy's signature problem – the scarcity of resources available for living a truly meaningful life. It is the long journey that takes us from the early population theories of Condorcet and Malthus to the evolutionary theories of Lamarck and Darwin, but which always circles back to eugenics (Fuller 2006b: chap. 13; Fuller 2007a: chap. 3). The last two sections of this chapter explore two latter-day manifestations of this polarity – between, on the one hand, a well-bounded species-type conception of the body politic that characterised the New Left and, on the other, exobiology's conception of intelligent life as a population diffused throughout the universe.

7 One step back to Weimar: The New Left's retrenchment of human embodiment

In 1976 I entered Columbia University as the product of a Jesuit scholarship school. I learned only later that my presence was part of a deliberate – and successful – shift in Columbia's recruitment strategy away from Jews. This was based on survey research that revealed Catholic students to be indifferent or hostile to the 'New Left', the omnibus name for the student-led, campus-based revolts against the liberal welfare states of North America and Europe during the Cold War – which also happened to sport prominent Jewish leaders. Nowadays the New Left is often presented as the source of academic postmodernism, especially when one wishes to create some ideological distance. Thus, a Richard Rorty (1998) or a Russell Jacoby (1987) nostalgically distinguishes between the righteous 'Old Left' of organised labour and the New Deal from the decadent 'New Left' of identity politics and political correctness. However, this sharp sense of an 'Old' and 'New' Left came into vogue only after the fall of the Berlin Wall in 1989. In its 1970 heyday, the New Left's characteristic shouts, obscenities, sit-ins, and interruptions of lectures was interpreted, even by mainstream academic observers, as signalling the need to re-think the distinction between rhetoric and violence (e.g. Johnstone 1971). Many of these observers were surprised at the ease with which the symbolic violence of the student activities was met by the authorisation by university administrators of the literal violence of police night sticks.

Moreover, as the 1970s wore on, academics appealed to the New Left's non-standard modes of rhetorical expression as a base from which to expand the concept of argumentation and communication more generally (e.g. Cox and Willard 1982). Nevertheless, they continued to wonder whether – and why – the New Left abandoned rational argument as an appropriate mode of rhetorical expression for dealing with the establishment. In this context, glib appeals to the philosopher Ludwig Wittgenstein were fashionable, especially his idea that language primarily 'shows' how the world is, a thesis taken to mean that language stimulates immediate intuition rather than discursive reason. Perhaps the communication theorists were overly impressed by the New Left's 'hippy' self-presentation – so much so that they failed to countenance that there might be more to their rhetoric than the 'safety valve' theory of criticism (as in 'blowing off steam') as instant gratification. Here it would have worth paying close attention to the 1962 Port Huron Statement, the rhetoric of which was partly inspired by the anti-establishment sociologist C. Wright Mills, who had prematurely died that year.

The Port Huron Statement, named for its Michigan university origins, served as the founding document of Students for a Democratic Society (SDS), the most thoughtful and best organised US student protest movement. The document's signatories present themselves not as outcasts but inheritors of what their parents have been promoting as the most politically and economically progressive society the world had ever known. The students realised that they were part of the first generation in which more than half of their cohort would receive a university education, which in turn would enable them to complete the 200-year American experiment by removing the last class, race and gender barriers to a prosperous, free and equal society. However, the advent of the Cold War threatened to pervert the outcome of this experiment, as the US government declared itself to be in a global ideological struggle of indeterminate scope and duration with the Soviet Union. In the balance lay the preservation of the 'American Way of Life'. Because these matters were being couched exclusively in military terms, there was a pretext for the concentration of decision-making authority in the chief executive, which was likely in the long term to erode the very civil liberties that the national defence strategy was supposedly designed to protect. The absurdist political theatre characteristic of New Left anti-establishment demonstrations is best understood as a play on this paradox.

The anchoring effect of the Port Huron Statement on the subsequent actions of the SDS should not be underestimated. It was published in the midst of John Kennedy's brief term of office, by which time he had already committed an unprecedented number of US troops to Vietnam and nearly started World War III with the failed Bay of Pigs operation in Cuba. All of this was authorised by the man whose 1960 presidential campaign had evoked much of the rhetoric on which the Port Huron Statement continued to draw. While the rhetoric probably reflected genuine sentiment and intent, it is equally clear that Kennedy and his overwhelmingly academic advisors – 'the best and the brightest', as the journalist David Halberstam scathingly dubbed them – operated on the assumption that one cannot maintain a democratic society without maintaining the appropriate geopolitical conditions, not least the removal of outright threats to the society's existence. If this means the suspension of some civil liberties for the duration of the struggle, then the end justifies the means.

Civic republican constitutions have always been tested on this point, to which they have formally responded by allowing power to be temporarily concentrated in an executive officer, called 'President' or 'Dictator', in times of national emergency who speaks and acts for all. However, the stress here supposed to be placed on 'temporary', implying that the threat is sufficiently well-defined that its presence or absence can be determined at any given moment. Indeed, declarations of war and other acts of aggression from recognised political units like states typically have an unmistakable beginning and end. More generally, this clarity of boundaries formed the backdrop of the civic republican *mythos*, which begins with a revolutionary moment, when a subjugated people join together to overthrow a tyrannical regime, after which they commit to the principle that their individual liberty in normal times is predicated on common purpose in times of need. The Second Amendment of the US Constitution enshrines this idea: The right to bear arms complements the duty to defend the republic. The problem is that the Cold War did not come with the moments of this *mythos* so neatly labelled. Indeed, the point of this 'war' seemed to be to avoid fighting potential aggressors, and many of them were thought to be already living within one's own borders as spies and traitors.

Admittedly, this paradigm shift in geopolitics was not fully appreciated at the time. Rather, the very name 'Cold War' suggested a vir-tual conflict between the US (and its allies) and the Soviet Union (and its allies), with the stress on 'virtual'. From that standpoint, the most salient aspect of the Cold War was its second-order – and hence second-guessing – quality, which drew attention to the then-innovative capacity of computer simulations as means for constructing scenarios, which in turn informed the policy of manufacturing arms for purposes of pre-empting or counteracting similar efforts presumed to be mounted by the enemy. Without denying the significance of these developments, which inspired the late great philosopher of hyperreality, Jean Baudrillard, to declare (more or less) that the Gulf War of 1990–1 did not happen, the failure of US Defence Department Secretary Donald Rumsfeld's 'shock and awe' strategy in the 2003 Iraq invasion should be enough to reveal the limitations of taking the lessons of strategic virtualism too literally.

The so-called War on Terror declared in the wake of 9/11 has painfully drawn a rather different precedent from the Cold War, namely, the foe's spatio-temporal elusiveness. What may turn out to have been most salient about the allegedly free-roaming Communist menace in the Cold War was not its official endorsement by the Soviet Union or the People's Republic of China but the relative ineffectuality of these two superpowers in directing Communism's spread, as the movement's worldwide devotees often operated in ways that placed the patrons at loggerheads. In this respect, insufficient attention has been paid to the ongoing tensions between Russia and China caused by the rather independent directions in which their supposed 'satellites' and 'client-states' took the Communist project. In retrospect, then, the Cold War may have marked the triumph of the parasites at the expense of the hosts. At least this has been the lesson learned by the Islamic terrorist cells operating from India and Pakistan to Britain and Germany, in which the (alleged) host nation-states have been left rather perplexed and unsure about an appropriate course of action.

This lesson has wreaked havoc on the civic republican self-understanding of nation-states, *especially* in the developed West, which has yet to acquire the cynical attitude to national sovereignty found in other parts of the world. Indicative of our continuing lack of cynicism is the instinctive revulsion – at least outside radical libertarian circles – to the idea of 'private security forces' (aka mercenaries), the civic republican's homegrown version of 'terror for export'. Indeed, this perhaps now naïve belief in the virtues of territorial sovereignty is one of many principles (including a commitment to a welfare state) uniting the Cold War establishment and their New Left opponents that have come under increasing criticism with the end of the Cold War. Where the two sides differed was over the trade-off between the exercise and the protection of nationally insured civil liberties. The New Left believed that the establishment's exaggeration of the Communist threat merely promoted the suspension of civil liberties without substantively protecting them. It should thus come as no surprise that leaders of the '68 student revolts like Joschke Fischer (Germany), Daniel Cohn-Bendit (France) and Tom Hayden (USA) later became professional politicians. They were revolting only against the perceived betrayal of civic republicanism, not the ideal itself.

The student radicals had been reading the Frankfurt School, especially Herbert Marcuse who by 1968 was already a well-known public intellectual in the United States. However, more important than the Frankfurt School itself to understanding the New Left's original rhetorical challenge is what the Frankfurt School was a response to, namely, the palpable failure of the Weimar Republic. Although the New Left has increasingly focussed on what might be broadly called 'cultural' and even 'lifestyle' politics, its deeper theorists have always kept a steady eye on what might be learned from the Weimar experience, in particular how changing material circumstances alter the terms in which the civic republican ideal is defended. Thus, Paul Piccone, founding editor of one of the academic journals spawned by the New Left, *Telos*, gradually shifted his concern from the original Frankfurt School to Carl Schmitt, the jurist who helped draft the Weimar Constitution but came to sympathise with the Nazi appropriation of its Article 48, the state of emergency clause, once the party was elected to form a government (Piccone and Ulmen 2002).

The Weimar hinterland of the New Left is worth lingering over. The Cold War conjured up the prospect of a Nazi-style 'permanent state of emergency'. Schmitt theorised this state as responding to the Achilles heel of civic republicanism, namely, its excessive reliance on the distinction between the power of the law (*potestas*) and the power of the lawgiver (*auctoritas*). It underwrote James Harrington's famous 17th century definition of a republic, 'an empire of laws not men', which so influenced the US founding fathers. The distinction had been an innovation of late medieval Roman law, attributable to John Duns Scotus' rather subtle detachment of divine power from divine paternity in the theological analysis of God (Brague 2007: 236–7). The profound secular implication, only fully realised with the Protestant Reformation, was that secular polities need not simulate biblical paternalism in order to retain their legitimacy. That we most readily understand God's power over us in terms of his creative capacity does not mean that the latter is required to legitimise the former. Similarly, the fact that a constitution required some specific person(s) for its original enactment does not bestow a power of enforcement on those person(s) or their descendants.

For Schmitt (1996), this devolution of power from paternity that over the previous 300 years had succeeded in undermining certain interpretations of the Catholic papacy, the divine right of kings, and hereditary rule more generally, now proved a nightmare in modern republics, where at least every elected official – if not every citizen – is entitled to an opinion on the appropriate deployment of the legally constituted *potestas*. It resulted in the procedural gridlock of parliamentary debate that Schmitt witnessed in the Weimar Republic, where no legislation could be enacted without a coalition of parties, each of which imposed its own conditions for cooperation, virtually ensuring that whatever was agreed would have limited efficacy. Schmitt's reactionary response was, in Max Weber's terms, a 're-enchantment' of the law through an extension of the state of emergency clause that would literally re-incorporate the republic in a single executive authority, the Führer, who would be empowered to act on behalf of all for as long as it took to remove all external threats. This conversion of the exception to the norm entailed a sense of 'materialist rhetoric' whereby a universal ideal was embodied in a concrete individual: Anyone who refused to follow the Führer was re-defined as an enemy of the people. What had been a matter of contesting interpretations of common ideals became as a life-and-death struggle against contaminants of the body politic.

Against this backdrop of the dark side of civic republicanism, the New Left may be seen as calling the old liberal establishment's bluff in its claim to both protect and permit free debate within the confines of 'national security', an expression whose scope in the 1960s appeared to increase on a daily basis. In this context, the 'street theatre' and 'antics' associated with the New Left's demonstrations – e.g. taunting police and administrators, commandeering cameras, sit-ins and other blockages of transit and communication – were designed to draw attention to power's brute physical presence. Even if not formally deployed, the levels of force and surveillance present at these demonstrations were of the sort also used to identify political foes.

Following up Schmitt's lead from an angle somewhat sympathetic to the New Left, the Italian social theorist Giorgio Agamben (1998) has highlighted a shift in the status of *Homo sapiens* in this state of emergency from *bios* to *zoe* – that is, from autonomous organism to mere living matter. Agamben's point is that we come to be treated less as purposeful beings than as sheer physical bodies whose vital signs are in need of regulation. Not surprisingly, perhaps, 'biology' was coined in the early 19th century by Jean-Baptiste Lamarck, who ascribed purposefulness to evolution, as opposed to, say, Charles Darwin, who generally avoided any reference to 'biology' when portraying what he regarded as life's relatively limited creativity in the face of natural selection.

Agamben's theorising acquired concrete purchase once George W. Bush unilaterally declared a War on Terror after 9/11, which has resulted in a preoccupation, most notably in the USA and UK, with gathering people's biometric data to keep their movements under surveillance. In this way, even ordinary citizens have come to be understood as 'always already' potential threats to the body politic. Sensitised by the Nazi and the Cold War precedents for this turn, Agamben drew the line in his own case by refusing to enter the United States in 2004 when he realised that his identity would be forced to undergo a transformation from *bios* to *zoe* at passport control (Agamben 2005). Shortly thereafter, in 2006, Students for a Democratic Society was reactivated largely in response to what Agamben had detected as a renewed attempt to undermine republican liberty in the name of protecting it.

The rhetorical take-home lesson here is that free and open rational argument requires specific background conditions to be in place. These conditions serve to de-materialise language, enabling all interlocutors symmetrical access to each other's knowledge claims, such that everyone enjoys the 'right to be wrong', and one does not feel compelled to stick to the same beliefs indefinitely (Fuller 2000a). Under such conditions, whatever people say is not held against them. However, when the republic undergoes a state of emergency, language turns into an extension of the law, which in turn has been re-incorporated in the sovereign, whose *potestas* then returns to the primordial state of *auctoritas*. In that case, language, even if expressed in declarative sentences, is effectively stripped of its conditional, subjunctive and even indicative modalities so as to be rendered purely imperative and injunctive. In short, language becomes an instrument of force, to which one responds with either submission or counter-force, the latter explaining the character of New Left rhetoric.

Here it is worth observing that this conversion of *bios* to *zoe* also occurs in what is often taken to be a positive legacy of the Weimar Republic to the New Left – namely, the legal protection of 'seriously held beliefs' that may require special dispensation from certain aspects of normal social life that would otherwise inhibit the free expression of those beliefs. Today, at least in Europe, this concern is most frequently raised in the context of Muslim customs, though if enforced the legal principle involved could make it as far-reaching as it was in Weimar (McVeigh 2009). This embodied pluralism – popularised nowadays as 'identity politics' – was originally enshrined in Article 118 of the Weimar Constitution as an extension of the principle of 'minority rights', the classic liberal safeguard against oppressive minorities. In *On the Essence and Value of Democracy* (1929), the great Viennese jurist and legal positivist Hans Kelsen justified this move without quite seeing its full practical consequences. In the Weimar period, 'minority rights' were still normally understood in ethnic terms but this was also the time when feminism and vegetarianism – lifestyle ideologies not reducible to class or ethnic markers – started to gain recognition, not least through the mass media and public demonstrations, both of which were Weimar social innovations. However, a pernicious long-term effect of this way of thinking about freedom of expression was the encouragement of fixed social identities for purposes of personal and political leverage. Of course, especially in cases involving ethnic identity, moves of this sort could be easily turned against those who made them, perhaps even providing grounds for segregation, deportation and – as the Nazis showed – extermination. In that case, we are back full circle to Schmitt's state of permanent emergency.

The above analysis also casts a somewhat unfavourable light on the perennial rhetorical strategy of speakers and audiences mutually 'adapting' to each other's assumptions when presenting arguments. The strategy sounds perfectly reasonable if each regards the other as a locus of attitudes and aspirations on which one can build and draw out in a mutually beneficial direction. That would be to treat interlocutor as *bios*. However, if one side already treats the other side as a potential threat to their existence, as *zoe*, then 'adaptation' takes on a more dysfunctional character that can be understood, to recall the Cold War lingo, as 'feedback loops'.

On the one hand, one may back down from the appearance of threat, perhaps thereby suppressing one's dialectical differences altogether. This is the conformist way of negative feedback – how the establishment liberals would have liked the New Left to respond. On the other hand, one may turn a mirror to the opponent and reveal that it is they who – at least equally, if not primarily – pose the threat, as in the infamous Pogo comic strip caption from Earth Day 1971, 'We have met the enemy, and he is us!' This is the confrontational way of positive feedback – as the New Left responded in its more violent moods. Perhaps the most constructive outcome of the New Left's positive feedback strategy was the renaissance in investigative journalism, the mirror image of the espionage carried out by the US intelligence agencies against perceived domestic threats, which reached its climax in the Watergate scandal, when *The Washington Post*'s Bob Woodward relied on the insider informant 'Deep Throat', who is now known to have been Mark Felt, the FBI deputy director whose agency was integral to surveillance of the New Left's activities. While both the negative and the positive feedback strategies were clearly 'adaptive' in their own way, neither held out the prospect of genuine evolutionary development. At best, a regime governed by negative feedback would restore to equilibrium the world that otherwise positive feedback would send to self-destruction – which explains the persistent anxieties of the more thoughtful Cold War cyberneticians, such as Norbert Wiener and Gregory Bateson (Heims 1991).

As the Port Huron Statement had already made clear, the New Left differed from the Old Left in believing that American society had reached a level of wealth and material security that entitled its members to political democracy. In contrast, the Old Left, be it in its revolutionary or reformist mode, still regarded most of the population as politically disenfranchised. Thus, their political aims veered between complete replacement of the *ancien regime* and gradual assimilation into the established order – that is, the socialist-liberal divide informing Cold War politics in the West. Because members of the Old Left did not yet see themselves as the equal of those who formally wielded power, they did not dedicate themselves so single-mindedly to the signature New Left tactic of re-appropriating the weapons of the power-mongers, whereby whatever means were used to reveal the establishment's own brutality.

8 Two steps beyond Darwin: Disembodying humanity in the search for extraterrestrial life

Charles Darwin's theory of evolution by natural selection, which celebrated its 150th birthday in 2009, is an interesting embarrassment for geographers and anyone concerned with time-space binding. Today evolutionists, including those who have taken holy orders (e.g. Ayala 2007), happily repeat Freud's claim in his *Introductory Lectures in Psychoanalysis* that Darwin followed Copernicus in displacing humanity from its collective narcissism. (Freud, of course, pretended to Darwin's mantle in carrying on the process.) To be sure, the claim has a certain *prima facie* plausibility, since for Darwin *Homo sapiens* is simply the latest moment in biological evolution – not the most advanced stage, let alone its final resting point. But on closer inspection, even on Freud's terms, the ascent of Darwinism is a case of two steps forward and (at least) one step back. Yes, Darwin provided a powerful empirical case that we differ from other animals only by degree, not kind. Thus, even on earthly terms, it is by no means clear that humans are the supreme creatures, especially given the cataclysmic transformations that have increasingly attended our dominion over the planet.

Yet, somewhat disconcertingly, Darwin brought science back to Earth, even though the Copernican revolution had been about displacing this planet's significance in the universe. In this respect, Newton is conveniently omitted from Freud's potted history of science. Even though the subject matter of physics and chemistry consists of elements and their combinations that may be found anywhere in the universe, the topics treated in biology remain largely localised to carbon-based forms that have descended – or perhaps can be extended – from those historically known to have populated Earth.

But there is nothing 'natural' about this reading of biology's scope. Pre-Darwinian biologists driven by monotheism thought of organisms as combinatorial possibilities that are realised only under certain conditions. Indeed, the original modern theorist of biological evolution, Jean-Baptiste Lamarck, portrayed God as given to trial-and-error, endlessly throwing forth possibilities to see if they take root in an inherently recalcitrant nature. But an even more obvious case in point is the greatest of all natural historians, excepting Darwin: Carolus Linnaeus, the special creationist responsible for the binomial nomenclature still used to classify species. His organisation of animal, vegetable and mineral kinds resembles the periodic table of elements, not the 'tree of life' that Ernst Haeckel popularised in the late 19th century to represent Darwin's theory of evolution, which continues to give 'the origin of life' *qua* ultimate ancestor such pride of place in biological thought.

This 'tree of life' image used to represent the lines of biological descent – or 'phylogeny' – on earth stems from August Schleicher's success in persuading Charles Darwin to extend it beyond its sketchy depiction in Chapter 4 of *Origin of Species* (Richards, R.J. 2002). Schleicher, Germany's leading comparative philologist at the time, introduced what is now called the 'cladogram' to chart the evolution of Indo-European languages from an original common ancestor in Northern India. The gradual exfoliation of this linguistic tree corresponded to migration patterns resulting in successive settlements farther in time and space from the 'Indo-Aryan' source.

The polyphonic rhetorical force of Schleicher's version of the tree image should not be underestimated. It kept alive Genesis-based ideas of a single and perhaps still recoverable root-language of 'being as such', in terms of which all subsequent languages are parochial elaborations and distortions. In secular guise, these ideas received their profoundest philosophical expression in Heidegger. Equally they provided justification for the idea of a 'homeland' (*Heimat*) associated with particular physical partitions of the linguistic migration, the basis of national identity, in terms of which one could speak of a further 'diaspora' that might be remedied through repatriation (e.g. Zionism and the resurrection of Hebrew). Finally the tree image could be used to measure progress in terms of distance from India, especially given the relative extremity of Britain, Germany and the Nordic countries. Indeed, this bias was already present in Hegel, who suggested a general East to West movement of the world-historic spirit to match the sun's passage over the earth. Hegel, who wrote in the first quarter of the 19th century, anticipated the United States as the ultimate resting place, with its mix of Germanic stock and English environment.

However, in the two decades since the 'Black Athena' thesis (Bernal 1987), which demonstrated the German philologists' systematic occlusion of Greece's debt to Africa, Schleicher's tree of linguistic transmission has been retired as one of the last vestiges of Eurocentrism – a century after one renegade philologist, Friedrich Nietzsche, had already begun to question the legitimacy of the search for origins altogether. The import of Nietzsche's 'genealogical' method was that a search back several generations might result in an ancestor similar in form but radically different in function, undercutting the idea that the mode of transmission had been 'meaningful'. Foucault subsequently underscored the point in his own 'archaeological' method, which also envisaged radical ruptures in forms, thereby subverting even the surface continuity displayed by the tree's exfoliated imagery. Nevertheless, the tree image remains alive and well in biology, even though it causes problems of a not unrelated sort when transferred to the concerns of Darwin and his successors (Atkinson and Gray 2005).

Take the doctrine of common descent – the idea that any two organisms share a more primitive ancestor, with all organisms ultimately descending from a single-celled creature. The doctrine is profoundly equivocal: Are we to believe that there is *exactly one* time, place and form from which all living things descended? Schleicher's cladogram certainly implied as much for the Indo-European languages. And while Schleicher was criticised in his own time for having failed to capture the descent of languages based in China (let alone regions outside Eurasia), his account was long accepted as reasonable for the Indo-European languages that have been the primary focus of linguistic inquiry. However, a phylogenetic tree, unlike its philological counterpart, charts biological change *only* in time, not space. It is as if once a new species emerges, either all its subsequent members, regardless of location, are traceable to that founder or the species is subject to multiple independent origins that happen to generate mutations of similar organisms under similar selective pressures. Yet, evidence is rarely, if ever, adduced for either of these astonishing hypotheses. When a project assuming these hypotheses was applied analogically to model human social history (Runciman 1989), the professional response ranged from indifference to scepticism.

Darwin's most prominent public defender of the past quarter century, Stephen Jay Gould (1988), provoked his colleagues endlessly by capitalising on the weakness of the cladogram as a map of evolution. He suggested that had the tape of Earth's history been rewound and replayed, a different configuration of species would have probably resulted. Gould's point, though presented much more genially, was rather like Nietzsche's: The survival and succession of forms – linguistic or biological – are local and contingent in all senses, including whatever resemblance the descendant bears to the ancestor. Thus, any surface unity or continuity presented in a cladogram is bound to be illusory. Invoking Darwin's authority in both cases, Nietzsche and Gould made their point similarly by noting the vast majority of species or proto-species that have become extinct over the centuries, amounting to an enormous waste of life. Under slightly different selective pressures, these life forms could have flourished and provided the genetic foundations for a set of descendants, whose unrecognisability would reflect, amongst other things, that we would probably not be around to recognise them.

This sense of contingency is covered up by the question-begging nature of Neo-Darwinian Newspeak. Consider the so-called 'molecular clock' hypothesis, whereby the time elapsed from when two species diverged from a common ancestor is measured in terms of DNA differences, themselves taken to be products of mutations that have been naturally selected. It follows that greater DNA differences mean earlier evolutionary divergence. For those who come to evolution by way of the laboratory, the molecular clock hypothesis provides an excuse to ignore fieldwork, the results of which – via the radiometric dating of fossils – provide new information only about the *timing* but not the *ordering* of species in the great phylogenetic tree, whose structure can be discerned from DNA alone (Watson 2003: 238–41).

In this context, the phrase 'fossil gene' has been coined to refer *not* to the genetic material left on the fossils of extinct organisms but to the genetic material present in extant organisms whose lack of adaptive value is presumed to be the remnant of an evolutionary ancestor (Carroll 2006). Needless to say, palaeontologists and field biologists see themselves as testing, not merely timing, the molecular clock hypothesis by the actual fossil record. One especially principled school, the cladists, refuses to postulate a common ancestor for two species unless there is concrete evidence that such a creature actually existed. Indeed, the cladist sees the 'tree of life' full of empty branches in ways that continue to fuel the hopes of special creationists (Gee 2000).

But having said all that, this quite justifiable scepticism about the validity of the tree of life image need not invalidate the prospects for biological science altogether. A universal life science need not be committed to Earth's natural history as its template any more than a universal social science need be committed to the history of Britain as its template. (However, the comparison both illuminates and casts an unflattering light on Marx's unrequited admiration of Darwin.) Instead one would want to know the general principles for creating and sustaining life that may exist *anywhere*. This project is not so strange if we focus on the dominant strand of the history of biology in the 20th century, which developed with little reference to Darwin's theory of evolution – from Mendelian genetics to the establishment of molecular biology, the DNA revolution, the mapping and sequencing of genomes, their systematisation and application as bioinformatics, alongside biotechnology-related developments involving the cyborganisation, xenotransplantation and the *in vitro* cultivation and synthesis of life forms.

For these fields nothing much hangs on the historical fact of whether animals and plants naturally evolved or were specially created, let alone whether it happened over 5000 or five billion years – least of all whether any of this had a unique time and place of origin. Indeed, excessive focus on these historical matters is potentially diversionary. After all, as we get better at mixing and matching strands of DNA, the origins of species may prove irrelevant to how any of them will survive in environments over which we will have increasing responsibility – if not exactly control. The general trend of this non-Darwinian side of biology is towards a conflation of ideas that Gould was especially keen to keep distinct: *homology of forms* and *analogy of functions*. In particular, the former is increasingly presumed to provide evidence for the latter. In other words, the fact that similar looking parts of organisms possess distinct evolutionary histories is overridden by a long-term tendency for those parts to converge on common functions. This has been historically the most persistent argument mounted against Darwin's indefinitely branching tree of life model of evolution, with its studied refusal to read a designer's handiwork back into nature.

Over the past 150 years, the anti-Darwinian argument has been successively mounted by St George Mivart, Pierre Teilhard de Chardin and, most recently, Simon Conway Morris (2003). These thinkers, all theologically inspired biologists, have regarded the advancement of science – including its results that contradict religious orthodoxy – as vindicating a progressive account of evolution culminating in humanity. Moreover, there are biologists who hold a watered down version of this view and call themselves Darwinists at the same time. They typically replace 'progressiveness' with something a bit less overtly purposeful like 'complexification'. Richard Dawkins springs to mind, especially given his weakness for teleological turns of phrase like 'selfish gene', 'blind watchmaker' and 'design without a designer'. Philosophers of science sometimes derisively dub Dawkins et al. 'adaptationists', suggesting that for them natural selection is simply a depersonalised version of 'Mother Nature', or what the old intelligent design theorist William Paley would have called 'The Mind of God'.

Nevertheless, as biology has moved from the field to the lab, and the training of biologists has come to resemble that of physicists and chemists rather than naturalists, design-based thinking has slowly reasserted itself – and along with it the idea of a universal science of life. Why should this be the case? The short answer is that biology is shifting from being a spectator sport to a creative enterprise: In theological terms, we are making the transition from animals to gods, a proportion that is explored in the rest of this book. It will seem blasphemous (to the devout), absurd (to the non-believer) or just scary (to the uncommitted). Here I would simply stress the word 'transition'. *Pace* Ray Kurzweil (2005) and his trans-humanist fans, I do not believe that 'the singularity is near' – at least not just yet. However, we are beginning to lay down the necessary social-epistemological infrastructure to make it a live possibility in the future.

One arena where conflicts relating to this transition are likely to be played out in the coming years is the prospect for extraterrestrial life, or 'exobiology' (Basalla 2006: chap. 10). This is the only topic that over the past half-century has led prominent physicists and biologists to question each other's scientific competence. Generally speaking, biologists are much less open than physicists to exobiology. Biologists quite reasonably argue that we have no reason to think that anthropoid intelligence would have proven adaptive elsewhere in the universe. But physicists do not think that settles the matter. For them 'anthropoid intelligence' is simply the name given for intelligence as it has emerged over the course of Earth's evolutionary history. From their standpoint, biologists reify that history. As physicists see it, there is no more reason to think that anthropoids are the only intelligent life form with which we could communicate than to think that we could recognise and trade only with nations that have undergone some version of Britain's political and economic history. Indeed, consider Britain's diminished yet respectable position as quite un-British versions of democracy and capitalism have spread throughout the world. But perhaps more to the point, physicists are prone to conflate exobiology with the search for extraterrestrial *intelligence* (aka SETI), where the electromagnetic bandwidth presumed necessary for conveying intelligence does not require a carbon-based container such as an earth-bound organism (Davies 2010).

Where physicists see differences in degree between ourselves and extraterrestrial life forms, in terms of which we may be outshone by aliens, biologists see differences in kind that preclude the very possibility of mutual recognition, let alone meaningful communication. Biologists have traditionally held the upper hand in the argument, given two earth-bound problems in getting the physicists' arguments off the ground. First, there are clear limits in our ability to infer higher mental powers from non-human species. Second, there is the so-called Weismann Doctrine, whereby, under normal circumstances, changes in an organism's constitution in its lifetime do not affect the constitution of the organism's offspring. The two problems have been traditionally connected together:

The cognitive impenetrability of other species supposedly reflects the constraints of 'vertical gene transfer', that is, the exclusive passage of genetic information along consanguineous, as opposed to, say, contagious lines. Bluntly put, we can only understand species with which we share an evolutionary history, regardless of the extent of our contact with them.

However, the burgeoning 'adaptationist' field of evolutionary psychology has unwittingly put pressure on the above restrictions. As researchers claim to find ever more similarities between the minds of humans and non-human animals, they are struggling to find ancient antecedents for those shared properties, so as to demonstrate the required common evolutionary history. Thus, claims about our 'reptilian' brains are meant to show not only the evolutionarily entrenched character of certain human responses but also the success of today's evolutionary psychologists in fathoming the reptile mind. Much of this work betrays a desperation one associates with astrology's many ingenious attempts to demonstrate action-at-a-distance, but at least it displays the admirable consistency with which Darwin's heirs defend the doctrine of common descent (Fuller 2008a: chap. 4).

Nevertheless, it is reasonable to wonder whether our emerging biotechnological capacities will ultimately make a signature Darwinian doctrine like common descent beside the point, when it comes to explaining, predicting and perhaps even understanding the behaviour of life forms. As biology's centre of gravity has moved from the field to the lab, so too our image of genetic material has shifted from that of seed (that contains all possible future realisations) to that of building block (that can be combined to produce any number of stable forms under the right conditions). Explanations of human behaviour in terms of our 'reptilian brains' seem powerful because we continue to think there is something to the idea that the present recapitulates the past, or 'ontogeny recapitulates phylogeny', in Haeckel's notorious phrase (which Nietzsche, perhaps with a nod to Schleicher, re-spun as 'ontology recapitulates philology').

But once that vestige of historicism is relinquished, any truth contained in the idea that we possess reptilian traits may be captured simply by the extent of contemporary human-reptile interactions, both socially and biologically. On the one hand, we are as 'reptilian' as reptiles can be incorporated into our life-world, which itself expands through greater interspecies interaction. (The same already applies for our judgements of disabled *Homo sapiens* as 'humans'.) On the other, our reptilian nature may be assessed by the replaceability of human parts, especially genetic clusters, with reptilian ones via xenotransplantation – and perhaps vice versa?

There are multiple signs of this gradual decline in descent as a biological concept. Some point to a time in the near future when our paradigmatic image of biological inheritance will shift from vertical to *horizontal* gene transfer, which spontaneously happens in nature when, say, a virus carries genetic material that spreads a disease across species boundaries (Dyson 2007). But of course much of biotechnology is also about the sorts of horizontal gene transfers – this time mostly deliberate – that occur during, say, xenotransplantation. Indeed, when/if we get to a point in which people can regularly reorient their genetic make-up by self-applied interventions, we will have turned the tide of evolutionary thinking from Darwin back to Lamarck.

Meanwhile, alongside this increase in our ability to redistribute genetic material across species boundaries has come a new vision of natural history that would turn the 'tree of life' image into an eccentric tendency in evolutionary time and space. Here I refer to the work of the US microbiologist, Carl Woese, who in the 1970s introduced the idea of a domain of life, whose single-celled members exchange genetic information amongst themselves and with other creatures by contact without forming clear species-like lines of descent. Woese called this new domain the *Archaea* because he believed it to be the most primitive form of life, which, just as Lamarck had originally imagined, continues to generate spontaneously under extreme temperatures (e.g. Woese 1998, 2004). To be sure, biologists have been happier to grant Woese's discovery of a new domain of life than to heed his call for a radical reconfiguration of natural history (Harman 2008: chap. 17). Not surprisingly, opposition to these more radical claims has been fiercest amongst the field-based Darwinists who also oppose exobiology. One of Stephen Jay Gould's early patrons, the late Harvard systematist Ernst Mayr, loomed large in both categories.

However, beyond displacing Darwin and rehabilitating Lamarck, an even more intriguing prospect for those interested in the restoration of something deserving of the name 'human sciences' is that biology is only now – and ever so slowly and painfully – undergoing a revolution in thought and practice that the social sciences underwent one or two centuries ago. I refer here to the shift in political and economic legitimacy from the inheritance of property to the redistribution of income as the principal vehicle by which the artificial construct of the nation-state maintained social order (cf. Fuller 2006b: chap. 4). This transformation was mediated by an abstracted and universalised sense of 'capital', whose exchange relations were to be regulated by norms that looked towards the future rather than remaining beholden to past entitlements. The result was socialism's most positive legacy.

But it is worth recalling that socialism generated not only utopian visions but also archaic antecedents like Marx's 'primitive communism', which served the valuable rhetorical function of providing a naturalistic back story for what, under even the best conditions, amounted to a massive restructuring of people's lives. I imagine something similar will be said, say, a century from now about the role of Woese's *Archaea vis-à-vis* the horizontal gene transfers that characterise the post-Darwinian biotech age on which we are embarking. At that point, biology will have finally caught up with the social sciences. Human geography contributes to this progressive project with its recent and somewhat ironic rediscovery of 'nature' as something whose materiality and agency are unleashed by losing what would have been traditionally regarded as its essential 'naturalness'. Meanwhile it is crucial that social scientists do not capitulate to Darwinian bluster – be it from Steven Pinker or Jared Diamond – that would downplay, deny, if not outright reverse, the advances that the social sciences have already made in the name of humanity by adverting to a sense of naturalism whose natural home is the annals of geocentric thought.

2
Defining the Human: The Always Already – or Never to be – Object of the Social Sciences?

Chapter 1 canvassed several struggles related to the material and ideological constitution of humanity but throughout I have been assuming that the social sciences are, so to speak, the 'party of humanity' in the academy. While historically this has been the case, is it so anymore – and if so, is it likely to be in the future? Chapter 2 addresses this question by considering the stakes, especially as the clarity of the social sciences as a distinct body of knowledge is increasingly in question. Nearly half a century ago, Michel Foucault identified the contingencies that originally enabled and nowadays disable 'the human' as a stable object of study and governance. Seen through the long lens of intellectual history, humanity has been the site of a bipolar disorder that has divided cognitive and emotional attachments between God and animal at least since the medieval foundation of the university. Indeed, the differences between the university's founding Christian orders, the Franciscans and the Dominicans, have redounded through the centuries, resulting in what I call alternative 'mendicant modernities': on the one hand our reabsorption into nature and on the other our transcendence of nature. The remainder of this book is devoted mainly to the latter prospect, an initial survey of which I provide in an inventory of 'transhumanisms', blueprints for 'Humanity 2.0'.

1 Does the success of the social sciences depend on our humanity?

From one viewpoint, the social sciences have never been more successful, especially in terms of available research funding and student course demand. Moreover, certain social science methodologies, notably those related to game theory, rational choice theory, and actor-network theory have been used to model phenomena in the life sciences. This would suggest that the social sciences are extending their influence across disciplinary boundaries. Yet, at the same time, 'social science' is losing its salience as a brand name or market attractor. In more academic terms, the social sciences are losing their distinctiveness as a body of knowledge distinguishable from, on the one hand, the humanities and, on the other, the natural sciences. That distinction was epitomised in the idea of a 'universal humanity' as both a scientific object and a political project that was explicitly developed by German Idealism, French Positivism, and the Socialist movements of the 19th and 20th centuries. These movements are rightly seen as spawn of the Enlightenment but, at a deeper level, they represent the secularisation of the Christian salvation project.

Each movement challenged, on the one hand, the humanities by declaring equal interest in all of humanity (not only the elite contributors to the 'classics') and, on the other, the natural sciences by declaring a specific interest in humans (in terms of whom other beings are treated as a secondary consideration, if not outright means to human ends). The common idea is that no one can be fully human until everyone is fully human. This is usually expressed as an ideal of equality but it could be also expressed as an ideal of liberty: Each person must be able to recognise everyone else as equals. The autonomy wished for oneself must be also extended to all others. These sentiments join the Golden Rule to Kant, Hegel and Marx. It also animates the modern ideal of distributive justice, which refuses to accept the incorrigibility of the disparity of wealth and achievement 'naturally' found among humans. This chapter should be understood as an extension of Fuller (2006b), a call to revive this robust sense of social science under the rubric of a 'new sociological imagination'.

However, this call for renewal faces an uphill struggle, as amply documented in Baber (2009), a symposium devoted to Fuller (2006b). As against my 'all and only humans' approach of the social sciences, the humanities and natural sciences are rediscovering their common historic interest in *human nature*, with stress now unequivocally placed on the 'nature' rather than the 'human'. In the face of the social sciences' tendency to attenuate if not outright reject human nature over the past quarter-millennium, the past 30 years have witnessed a steady stream of works purporting to 'unify knowledge', most explicitly Wilson (1998), which in practice would make direct links between the classical humanities and the modern natural sciences by circumventing the social sciences altogether in the name of 'human nature'. Here Darwin replaces Aristotle as the grand unifier. In this context, the concept of universal humanity and most social science *theories* (though, as I have already suggested, not social science methods or findings) appear as vestiges of a monotheistic worldview that would elevate the human condition above the rest of nature.

Notwithstanding the radically different biologies that underwrote their conceptions of human nature, Aristotle and Darwin both doubted that the traits most closely associated with normative conceptions of 'humanity' were equally distributed across all members of *Homo sapiens*. Whereas Aristotle and his contemporaries argued about the limits of pedagogy in converting the upright ape into a political animal, Darwin and his successors have suspected that the upright ape's various attempts to transcend its biological condition – be it via Christianity or Socialism – simply reflects a pathology in an overdeveloped cerebral cortex.

Moreover, the general prognosis of the re-absorption, if not outright 'withering away', of the social sciences into a broader conception of nature has been also advanced by a consensus of postmodern social theorists who have queried the ontological significance of the human/non-human distinction and the need for disciplinary boundaries altogether (Latour 1993; Wallerstein 1996). However, their anti-dualism is informed less by a desire to reduce the mental to the physical than to reveal the interpenetration of the two categories, such that spirituality or consciousness is no longer seen as unique to humans but common to even the simplest forms of matter. This is not behaviourism or even materialism, at least as conventionally understood by physics-minded philosophers, but something closer to hylozoism and even panpsychism (e.g. Deleuze and Guattari 1987). Such convergence between naturalists and postmodernists should be unsurprising, given their common basis in Darwin's explicitly non-teleological version of evolutionary theory. In the postmodern case, it is filtered through Nietzsche's 'genealogical method'. The benchmark text here is Michel Foucault's *The Order of Things* (1970), his most sustained 'archaeology of knowledge', which focussed on the sudden emergence and gradual disappearance of the object 'man', that 'empirical-transcendental doublet', the Kantian phrase that Foucault used to characterise the distinctive nature of our being. In the Foucaultian gaze, we are exotic apes suffering from what Richard Dawkins (2006) calls the 'God delusion'.

Foucault notoriously regarded humanity as a historically bounded object that really only came into existence with Kant's coinage of 'anthropology' in 1798 when he was addressing how beings of such diverse racial-cultural histories as those Linnaeus had canonised less than 50 years earlier as '*Homo sapiens*' could ever deliver on the Enlightenment promise of 'world-citizens' – or 'cosmopolitans', to recall the original Greek (cf. Toulmin 1990). Kant's cosmopolitan conundrum slowly began to lose its salience a hundred years later, as Marx, Nietzsche and Freud, each in his own way, portrayed the human as an unstable compound, a 'house divided against itself' subject to false consciousness, self-deception, and/or repression. For them 'humanity' in this grandiose sense merely encouraged people to live in an unrealisable future that diverted them from the intractable problems they currently faced.

Contrary to most of Foucault's critics, I accept the *prima facie* plausibility of his radically demystified account of the concept of humanity, which in turn demands a systematic response, one begun in Fuller (2006b). Foucault is certainly correct that a distinct body of knowledge called the 'human sciences' or 'moral sciences' or 'social sciences' that takes all human beings to be of equal epistemic interest and moral concern has been most compelling from the late 18th to the late 20th century. For Foucault himself, this was a blip on the radar of Western intellectual history, on either side of which he espied (before) an enchanted and (after) a disenchanted naturalism: in short, Aristotle and Darwin. Even those operating within a more conventional view of intellectual history can recognise Foucault's 'Age of Man' as signifying the shift from a broadly supernaturalist to a broadly naturalist worldview: For example, where once wars were fought about the right approach to God (theology), wars in the future are likely to be fought on the right approach to nature (ecology), with the familiar modern inter-state conflicts licensed by the Peace of Westphalia of 1648 functioning as an extended transitional phase between the two pure forms (sociology).

2 The precariousness of the human: Why Foucault is (unfortunately) correct

As propaedeutic to my own response, we need to get the full sense of humanity's ontological precariousness to appreciate the depth of Foucault's challenge. It is epitomised in the following question: *Have we always, sometimes or never been human?* The more one understands the history of the concept of humanity, the less the question appears frivolous. To take the question seriously, one should take into account the following four considerations:

1. *There has always been ambiguity about where to draw the line between humans and non-humans* (Corbey 2005). It can be found even in Linnaeus' coinage of our species name *Homo sapiens* in the mid-18th

century. At the level of morphology, Linnaeus did not see a sharp difference between the higher order apes and the various human races. However, as a special creationist of the Lutheran persuasion, Linnaeus believed the biblical claim that all humans were endowed with souls that gave them the potential to hear God's call – even very fallen humans, such as the sons of Ham cursed by Noah, from whom Africans were thought to descend. In this respect, the Bible made up for the shortcomings of empirical observation in providing a clear definition of the human (Koerner 1999). It is only a short step from this line of reasoning to the 'standard of civilisation' long enshrined in international law – that a people are properly 'human' only if they heed the call of God or at least, in more secular terms, tolerate the commerce of those who do (Fuller 1997: chap. 7). The behaviourist orientation of this approach is striking. The distinctive spirituality of the human is marked by one's responsiveness to a sacred book in which the distinction is itself inscribed. Thus, one reason why so many more American Indians than Black slaves were slaughtered in the United States was that the former refused to adjust their mode of being in response to the divine call.

2. *There has always been recognition of the diversity of physical and mental qualities of beings that might qualify as humans.* Sometimes the originality of this observation is credited to Darwin but only because folk notions of species tend to presume a crude understanding of essentialism. Even Aristotle knew that a species contains differentia: The same thing may exist in many different ways, amongst which exists what Wittgenstein called a 'family resemblance' that, in turn, points to a common ancestry. Where Aristotle and Darwin disagreed was that the former thought of this variation as resulting from a mixture of elements provided by the particular parents whereas the latter saw it as endemic to the general process of reproducing the species. Nevertheless, followers of both Aristotle and Darwin have had their doubts about the capacity of all members of *Homo sapiens* to achieve the same levels of humanity. To be sure, they believed that all members of the species possess a sufficiently joined up nervous system to merit the minimal infliction of gratuitous pain. But otherwise, people are inherently different, which means that the just society is organised by enabling each person to flourish in the sort of life that he or she has been designed to lead. In this respect, the division of labour is simply the outward sociological expression of a natural biological tendency, which (so at least Plato believed) philosophy could rationalise.

3. *There has always been an understanding that not everything about humans makes one human and that one's humanity might be improved by increasing or decreasing some of its natural properties.* In other words, to be human is to engage in activities whose purpose goes beyond the simple promotion and maintenance of the animal natures of those qualified to be human. In Western philosophy and theology, one normally characterises such matters as involving a 'spiritual' or 'intellectual' quest, but it is perhaps less misleadingly cast as a call to artifice. Here we might identify three 'Ages of Artifice'. (1) *The Ancient Artifice,* epitomised by the Greek ideal of *paideia,* instilled humanity through instruction on how to orient one's mind and comport one's body to justify one's existence to others as worthy of recognition and respect. In practice, this meant speaking and observing well – the source of the liberal arts disciplines. (2) *The Medieval Artifice,* epitomised by the introduction of *universitas* into Roman law, promoted humanity by defining collective projects into whose interests individuals are 'incorporated', say, by joining a city, guild, church, religious order or university. Here one exchanges a family-based identity for an identity whose significance transcends not only one's biological heritage but also one's own life. (3) *The Modern Artifice,* epitomised by the rise of engineering as a distinct profession, advances the human condition by redesigning the natural world – including our natural bodies – to enable the efficient expression of what we most value in ourselves.
4. *There has always been recognition that genuine humanity is precious and elusive, and hence 'projects' and 'disciplines' for its promotion and maintenance have been necessary.* In a sense, this is the negative side of humanity's inherent artificiality, noted above. It implies that the pursuit of humanity may not necessarily serve the interests of all flesh-and-blood humans. At the very least, not all humans may benefit to the same extent and in the same way from the process of 'humanisation'. The easiest way to appreciate this point is to consider what it would take to realise any of the historically proposed schemes that would establish 'equality' among all humans. Some individual humans would be raised and others diminished in the process, the balance between which would always need to be monitored. Christianity is largely responsible for inducing widespread cultural guilt about the failure of all members of *Homo sapiens* to be treated as humans, which in turn opened a long and ongoing discussion about how 'human potential ' (aka soul) might best be realised. However, it is only in the late 18th century that the first systematic efforts to raise 'the overall level of humanity' by the

redistribution of wealth and sentiment are instituted, this time in the name of 'Enlightenment' (Fleischacker 2004). In the 19th and 20th centuries, these efforts came to be routinised as a set of political expectations concerning mass education, health care, and welfare provision more generally.

The perceived failure, or at least underachievement, in securing the fourth sense of humanity's ontological precariousness has led even self-avowed members of the political left to judge 'humanity' a fantasy whose inherent risks are outweighed only by its manifest hypocrisy. Often Foucault's 'death of man' thesis underwrites this conclusion but as suggested by the five critiques listed below, this anti-humanism can be found in a variety of contemporary trends, some of which stray far beyond Foucault's original concerns but all of which are well represented in John Gray's (2002) *Straw Dogs*, the most provocative British book of political theory in recent times:

1. *The Postmodern Academic critique*: 'Humanity' is a mask that hegemonic male elites don to exert power over everyone – and everything – else.
2. *The Neo-liberal (and Neo-conservative) critique*: Humanity as a political project costs too much and delivers too little (aka race and gender are 'really real').
3. *The Ecological critique*: The projects associated with humanity are depleting natural resources, if not endangering the entire biosphere.
4. *The Animal Rights critique*: Humanity's self-privileging is based on pre-Darwinian theological ideas that cause other creatures needless suffering.
5. *The Posthumanist critique*: Not even humans want to associate with other humans any more – they prefer other animals and the 'second selves', or avatars, they can create on their computers.

Humanity's perennial precariousness may be appreciated upon considering that prior to the Stoics the classical philosophers probably did not count all members of the species *Homo sapiens* as 'human' in a normatively robust sense, whereby '*Homo sapiens*' names only humanity's contingent biological starting point but not its ultimate realisation. On the one hand, when Aristotle defined the human as *zoon politikon* ('political animal'), he seemed to be referring only to those with the capacity to participate in public life, i.e. male landholders in good social standing. On the other, a quarter-millennium later and under the rubric of *humanitas*, the Stoic Cicero commended a variety of orientations to the world that transcend ordinary brute survival. They included the cultivation of leisure as an end in itself (and not simply a respite between periods of work) and the recognition of both what others accomplish on their own and one's own dependency on others (thereby evening out the natural tendency towards pure self-interest). For Cicero, himself a semi-invalid provincial who eventually achieved greatness in the Roman Senate and as a writer of Latin prose, *humanitas* helped to explain his own success. Indeed, Cicero's contemporaries deemed him a *novus homo*, a 'self-made man', the ultimate compliment that could be paid to a being so marked by artifice. In this respect, *pedagogy*, the ancient discipline that grew out of rhetoric and preceded government as the means for radically and systematically amending the upright ape's default tendencies, should be seen as the low-tech precursor of the various treatments increasingly available today for 'cognitive enhancement' (cf. Ingold 1994; Harris 2007).

Here it is important to recall that in the classical world the default social unit was the family estate, or household (*oikos*), and *not* the city-state (*polis*). The latter only came into its own under extraordinary conditions, either in times that require mutual aid (i.e. in war or a famine) or when the material bases of life have been already served (i.e. in leisure). In contrast, the household was the natural habitat of several families of human and non-human species (i.e. farm animals) that have long co-existed symbiotically. It involved a functional differentiation of resource production and management based on the workings of 'natural justice', that is, the spontaneous variation in individual talents within a species. Just as one would not expect everyone to be equally capable of hard physical labour, the same would be true of the mental discipline necessary for becoming 'human' in the normatively robust sense indicated above. In this sense, a 'just society' removes any artifice that might prevent heredity from operating as an efficient sorting mechanism for assigning individuals to their appropriate societal functions. Thus, the less articulate would not feel the burden of having to speak 'rationally' because they would be valued for their other natural capacities. In short, the 'just society' did not denote a vehicle for collective self-improvement – as that phrase would come to mean in the modern era – but a sustainable ecology of mutually complementary individuals.

In this context, Aristotle was more trusting than Plato of natural justice, evidence for which can be found in Aristotle's rather charitable view of Athenian drama, which tended to feature plots in which one or more characters tries to act contrary to nature, only to fail in some comic or tragic way. A good contemporary exemplar of the Aristotelian attitude is the controversial US political scientist Charles Murray (2003), who has never ceased to find the comic and tragic elements in state-based welfare schemes. As for Plato, while he appreciated the persuasive force of hereditary appeals, especially as a socially stabilising ideology, in the end his scheme to recruit and train philosopher-kings was about finding the best individual for the job based on rational criteria, which justified a policy of artificial selection. From his standpoint, the Athenian dramatists ran interference on humanity's capacity to realise nature's ends more fully. However, when it came to the city-state as a society of self-legislating equals, both Plato and Aristotle found it an alluring but potentially self-destructive chimera.

What Plato and Aristotle lacked was a criterion of humanity that overcame the obvious morphological and behavioural differences among members of *Homo sapiens*. A century later, to the rescue came the rather complex Stoic idea of the soul, which encompassed words like *pneuma*, *psyche*, *conatus* and, most notably, *logos*. These words captured the source, the expression and the perpetuation of life. By today's standards the nature of these entities blurred distinctions between the psychological, biological and physical. Each was located somewhere between a meme and a gene, the sort of thing that only a Lamarckian could truly love. But together they served to shift the burden of proof to a recognisably universalistic notion of humanity, whereby instead of marvelling at how well certain people can speak in public (the canonical expression of *logos*), one wondered why everyone else *cannot* do so as well – given that God had also endowed them with a soul.

Perhaps Stoicism's most enduring legacy to the concept of humanity has been the Christian gloss on *logos* as divine agency in the Gospel of John. However, John radically shifted the metaphysical horizon of *logos*. The Stoics regarded humans as embedded in a pantheistic universe: Our capacity to resist the animal passions and to reason beyond our immediate needs reflected our unique status as a microcosm of the universe – but nothing more. In other words, as Spinoza continued to believe, humanity was simply the locus of God's self-understanding, where 'God' is simply a pious name for nature in its entirety. In contrast, the Christian God unequivocally transcends the world of his creation, and humans are defined as those created 'in his image and likeness', what after St Augustine has come to be known as the *imago dei* doctrine (Fuller 2008a: chap. 2). The difference between Catholic and Protestant sensibilities turns on what one takes to be the main feature of the *imago dei* doctrine: *our subordination* or *our likeness* to God

– what in the next section I characterise as, respectively, the 'Paris' and 'Oxford' spin on the doctrine.

3 Humanity as a bipolar disorder and the legacy of John Duns Scotus

Foucault understood well the latent source of anti-humanism in the Western tradition. Western theology poses the question of humanity in terms of whether we are more like gods or apes. From a sociological standpoint, either answer ends up devaluing what most normal human beings do, or at least what they believe about what they do. Those who would urge humanity's apotheosis are eager to discipline, replace, if not outright eliminate our animal natures to release a frictionless medium, typically of thought, that enables us to merge with God. Their sense of science is ascetic, such that experiments function as trials of the soul, where the inquirer is pressed into extreme situations that challenge our physical senses to elicit a significant response (Noble 1997). The flagship discipline of apotheosis is *optics,* whose imprint is still felt in the hype surrounding superconductivity research and the eagerness with which people embrace avatars in cyberspace and speculative attempts to download consciousness into silicon chips. In contrast, those stressing our ontological proximity to the apes have tried to show the continuities in our natural modes of being with those of the rest of the animal kingdom, typically to parlay a respect for nature into a sense of humility, if not submission. Their sense of science involves full sensory immersion in a habitat, the flagship discipline of which is *natural history*. From this standpoint, Darwinists overstep the line of theological respectability *only* when, as in the case of the 'new atheist' followers of Dawkins (2006), they infer the non-existence of a supernatural realm simply because it cannot be accessed through science: Rather, as the Roman Catholic Church has stressed since Thomas Aquinas, natural scientists should understand that other modes of being require other modes of knowing.

For today's version of the 'gods' *vs.* 'apes' poles that pulls apart the integrity of humanity, consider Ray Kurzweil's (1999) 'spiritual machines' and Peter Singer's (1999) 'animal liberation' as radically alternative 'post-human' ends – that is, what humans should be about, understood in terms of the larger reference group with which we wish to identify in the future. Speaking of 'human' as what Nelson Goodman (1954) would call a 'projectible predicate', we see here what I have called the great *carbon-silicon divide* in human being (Fuller 2007b: chap. 2). It is epitomised in the following question, which brings out the alternative modalities at stake: *Are we by nature intellects that happen for now to possess animal bodies (Kurzweil), or animals that happen for now to possess distinctive minds (Singer)?* The former option suggests a purposeful intelligence who explores different media for optimal self-expression, each disposable if proved unfit for purpose. The latter option implies a completely contingent process that reduces any noteworthy effects to emergent properties of particular combinations of elements, the valorisation of which is superstitious. Theologically speaking, Kurzweil and Singer are guilty of complementary excesses that recall the Gnostics and the pagans, respectively, as boundary challengers to Christianity from its earliest days. Kurzweil the Gnostic would 'sacrifice' his own body – and perhaps that of others – in service of immortal life, while Singer the pagan would 'sacrifice' his higher mental functions in the sense of removing the inhibitions they normally pose to a full return to our sensuous, mortal roots.

Notwithstanding the contemporary focus of their interests, Singer and Kurzweil are reproducing signature attitudes towards the ends of knowledge that are traceable to the 13th century university foundations of Paris and Oxford, respectively, and their contrasting views of the ends of humanity. This 'bipolar disorder' is captured in Table 2.1, which is informed by the excellent treatment of John Duns Scotus presented in Williams (2007). (See also Fuller 2008a: chap. 2). In this section and the next, I explore how this disorder comes to define competing versions of modernity, which in section five I dub *mendicant modernities*, as they carry forward the in sacred and secular guises the worldviews of the two great religious orders that staffed the early universities, the Dominicans and the Franciscans.

John Duns Scotus (1266–1308) is central to this discussion because *Homo sapiens* embarked on 'humanity' as a collective project of indefinite duration only once an appropriate metaphysical framework was in place to make good on the biblical idea that humans have been created 'in the image and likeness of God'. Although St Augustine first crystallised *imago dei* as a theological doctrine, it really only comes into its own a millennium later, with Duns Scotus, largely in response to the tendency by followers of Thomas Aquinas, or 'Thomists', to multiply realms of knowing and being, seemingly to check the ambitions of the will as the locus of humanity's God-given creativity. At issue here is what the medieval scholastics called 'divine predication', that is, the nature of the terms we use to speak of God's qualities.

When we say that God is 'all good', 'all powerful' and 'all knowing', are we using 'good', 'powerful' and 'knowing' in the same sense as when speaking about ourselves? Scotus said yes and Aquinas no. Thus, Scotus

Table 2.1 Humanity as a Bipolar Disorder: Paris *vs.* Oxford

UNIVERSITY	PARIS	OXFORD
RELIGIOUS ORDER	Dominican	Franciscan
ACADEMIC EXEMPLARS	Albertus Magnus, Thomas Aquinas	John Duns Scotus, William of Ockham
PHILOSOPHICAL ANCHOR	Aristotle	Plato
GOD IS ...	Simple	Infinite
'BEING' IS ...	Equivocal	Univocal
KEY ONTOLOGICAL DIVIDE IS ...	Between God and creatures	Between humans and animals
HUMANS ARE ...	High-grade creatures	Low-grade creators
GOD-HUMAN RELATIONSHIP	God and we are 'simple' in different senses, aka logical vs. psychological simplicity	We are 'finite' and God is 'infinite' in the same sense, i.e. in terms of extension
DIVINE ATTRIBUTES ...	Are endlessly additive but no convergence because our language cannot refer to God	Converge at the limit because our language (as *logos*) ultimately refers to God
A PRIORI KNOWLEDGE?	Only to the extent that God grants it	It is how we always participate in divine mind
EPISTEMIC CERTAINTY	We can achieve it only in the natural world	We can achieve it on our own (with God's help)
WILL IS ...	Intellectual appetite	Creative source of being
WILL IS BASED ON ...	*Affectio commodi* ('sense of advantage' – what best suits our interests)	*Affectio iustitiae* ('sense of justice' – what the part looks like from an imagined divine whole)
SYSTEMS PERSPECTIVE	Local adaptation	Global optimisation
WE ARE FREE TO DECIDE ...	Whether to do what is (already) right or wrong	Which path we take, which may have good or bad consequences
BASIS FOR ETHICS	To live a flourishing life	To explore human potential to the fullest
WHAT IS POSSIBLE?	Whatever is probable (i.e. an empirical notion)	Whatever is conceivable (i.e. a semantic notion)
WHAT IS PROGRESS?	Increasing differentiation, complexification	Increasing purification, demystification

argued for a 'univocal' theory of predication, Aquinas for an 'equivocal' (or 'analogical') one. Although it took another quarter-millennium for the full implications to sink in, the Scotist view made it possible to think of humanity along a continuum – 'great chain', if you will – of beings, on the far end of which stood God. Thus, the non-conformist Christians active in Europe's Scientific Revolution interpreted the Scotist doctrine of the 'univocity of being' – that divine qualities exceed human ones only in degree but not kind – to imply that we might not only understand and improve our animal existence but that we might even chart a path to divinity (Funkenstein 1986; Harrison 1998).

In terms of this trajectory, Scotus can be seen as having contributed to a quasi-mathematical understanding of the Platonic doctrine of *participation*, whereby at least part of our being overlaps with that of the deity in terms of our ability to grasp the eternal forms that serve as templates for material reality. Thus, after Scotus, one could easily imagine degrees of participation in or, as his fellow Franciscan Bonaventure – the great 13th century rival of Aquinas – originally put it, distance from reunion with 'the mind of God'. This Platonic doctrine had been strongly represented in the Muslim philosophical tradition (Avicenna, Averroes, etc.) prior to its re-introduction into Christian Europe, since it suited the orthodox construal of Muhammad's role in the production of the Qur'an, namely, as a willing and pure vessel of divine transmission who was not compelled to compromise his material integrity as a human being. This suggested that our rational faculties were divinely inspired and capable of indefinite expansion. The Franciscans made this point explicit, but it only started to be understood in more materialistic terms – e.g. that we might exercise dominion in ways that successively approximated divine governance – in the Scientific Revolution.

'Humanity' thus became the name of the project by which *Homo sapiens* as an entire species engaged in this ontological self-transformation. The idea implied here, *achievable perfection*, is the source of modern notions of progress, ranging from scientific realist attitudes that 'the true' is something on which all sincere inquirers ultimately converge (regardless of theoretical starting point) to more general notions of progress, whereby 'the true' itself is held to converge ultimately with 'the good', 'the just', 'the beautiful', etc. in some utopian social order (Passmore 1970).

At this point, it is customary to cite Kant's *Critique of Pure Reason* (1781), published as the Enlightenment was reaching its climax, which put the brakes on this line of thought and set the pace for an increasingly sceptical attitude towards human self-transcendence over the next two centuries. In his discussion of the 'antinomies of reason', Kant re-spun the Thomistic line about multiple modes of language to conclude that when the 'rationalists' of his day adopted a Scotist line on divine predication, they were extending ordinary usage beyond the bounds of sense, 'language on holiday', as Wittgenstein would later archly put it. These rationalists had one foot in science and the other in religion – and would be classified today as 'natural theologians'. According to Kant, they had become bewitched by language by presuming to speak about a being – God – whose existence could not be firmly established by our normal logico-empirical means. But whereas the Thomists concluded that one had to turn to the distinctive experience of faith at such moments when language fails, more secular thinkers in the 19th and 20th centuries – epitomised by the logical positivists – simply concluded that religious language is a strictly meaningless outburst of emotion.

While Kant's view certainly had a corrosive effect amongst philosophers and intellectuals, we need to return the Scotist shift of humanity's centre of gravity from animal to deity to see the source of the modern worldview. Consider a standard way of thinking about God as existing outside of space and time – i.e. *sub specie aeternitatis*. Thus, Newton explicated divine omniscience as a function of God being equidistant from all times and places. What appears to change within space and time appears from the 'view from nowhere' as always equally present (Nagel 1986). A measure of just how far short humans normally fall of this standard of divine knowing is our default asymmetrical attitudes towards the past and the future: We ordinarily believe that the past, even if not knowable in fact, is at least fixed in principle, while the future remains unknowable in principle because it has yet to be fixed in fact. What modernists call 'traditional', 'conservative' or 'pre-modern' societies, experience this asymmetry as especially significant: Traditionalists invoke the past to enforce the legitimacy of the present, while they portray the future as hazardous if it breaks definitively with the past. The modernist imperative, in contrast, has been to redress such asymmetrical attitudes towards past and future, so as to bring them closer to God's point-of-view, whereby what we call 'past' and 'future' are equally knowable – and (to God) known. In effect, modernity has tried to simulate the divine standpoint from a human position by arguing that our knowledge of the past is not so secure and our knowledge of the future not so insecure. In its own fallibilistic way this adjustment of perspective approximates Newton's divine ideal of being epistemically equidistant from all moments in space-time.

Perhaps the easiest way to see this strategy at work is to look at the dual movement of the modern scientific attitude towards temporal affairs. On the one hand, modern science is sceptical towards received views of the past, not least because of science's extended sense of time's backward reach, which serves to cast doubt on both the constancy and the reliability of the information that can be gathered about alleged events now thought to be so far from the present. This is the basis of the critical-historical method that was the hallmark of the 'humanities' once it started to name a set of academic disciplines in the early 19th century. It also undercut the basis for traditionalism, which was increasingly seen as a mythical construct. On the other hand, modern science takes a more positive view towards our capacity to know and even control the future, especially through predictive experiments. Here the idea that the 'end of time' is not near but in the indefinite future allows the prospect of collective learning through a controlled process of trial and error, what Karl Popper (1957) famously called 'piecemeal social engineering'. Thus, the future need not be faced with foreboding and fatalism but with openness to change, since there is time to make and correct mistakes, and indeed to refine our sense of long-term planning. This is what is typically meant by positive references to a 'natural science' attitude to human affairs that is already present in the works of 'utopian socialism'.

At a metaphysical level, the strategy to simulate a divine symmetry in humanity's attitudes to the past and future reflects the Scotist shift in the default sense of what is 'possible' from *the probable* to *the conceivable* (Fuller 2002b). This, in turn, had a knock-on effect on the understanding of human agency. On the one hand, since the past was no longer treated as such a secure guide to the future, the current generation of humans had to take personal responsibility for what to do – they could not simply defer uncritically to dead ancestors. On the other, since the future was no longer seen as so threatening, the current generation could play with alternative courses of action, perhaps comforted by the thought that future generations, with the benefit of both hindsight and insight (if not foresight), might be able to cope with the consequences better than those whose actions generated them. An important transitional figure was the British Unitarian preacher, experimental chemist and confidant of the US founding fathers, Joseph Priestley (1733–1804), who explicitly regarded scientific progress as the expression of divine providence (Passmore 1970: chap. 10). The totality of Priestley's corpus wedded a deep but deconstructive approach to the history of Christianity to an equally deep and productive commitment to experimental science and utopian politics (Johnson 2008). Arguably he was the first fully embodied 'modern man'. We shall return to Priestley in Chapter 4.

So far I have been interpreting the Scotist imperative largely in epistemological terms. Thus, the focus has been on our ability to simulate the divine standpoint, controlling for the ultimate difference that humans exist inside – and God outside – space and time. But the Scotist task may be understood in more strictly ontological terms that takes seriously the idea that humanity and divinity are two ends of a continuum of being. In that case, it matters from which end one starts to approach the other. On the one hand, approaching the divine from the human pole involves rendering the mortal immortal, such that God is defined as the being who never suffers death. On the other hand, approaching the human from the divine pole would involve perpetuating the eternal, such that God's timelessness appears as an indefinitely reproduced state of momentary creation. These two options are quite pointedly related to the person of Jesus as a being who somehow provides a precedent for all of humanity by travelling along this ontological continuum.

While it would be a mistake to see the modernist project of humanity as the straightforward outcome of Duns Scotus' subtle logical analysis of being, nevertheless Scotus sowed its seeds when he introduced the abstract conception of law as something universally binding and equally accessible. He did this by a linguistic innovation that effectively divested the force of law of its divine origins. As Brague (2007) observes, in Latin the shift was signified by the replacement of the proper name *pater omnipotens* ('almighty father') with the generic attribute *omnipotentia* ('omnipotence'). This stand alone conception of power opened the door for all of God's properties to be reconceptualised as dimensions for comparing the human and the divine, in terms of which one might speak of 'progress' from the former to the latter. A latter-day descendant of this profound shift in Western consciousness is the 'transhumanist' mentality that we shall encounter at the end of this chapter and throughout the next. It too tends to think of 'enhancement' in terms of the indefinite promotion of single virtues, so much so that, say, intellectual or physical prowess becomes a positional good (Hirsch 1976), that is, something whose value is defined primarily in terms of its relative scarcity. Thus, as long as everyone wants to be cleverer – say, to gain competitive advantage in the labour market – then one can never be too clever. In Chapter 3, I associate this development with the ideology of 'ableism' (Wolbring 2006).

However, the most immediate and sustained consequence of turning divine qualities into generic attributes is that social life came to be reorganised in specific ways that we continue to take for granted. I earlier referred to the 'Medieval Artifice' known as the *universitas*, whereby individuals would be legally incorporated into larger social wholes, like cities and universities, on a non-biological basis. Scotus' linguistic innovation was accompanied by a sensibility that what previously had been concentrated in God for eternity, and perhaps harnessed by his papal or royal representatives on Earth for a limited time, could be instead distributed over an indefinite period as an increasingly secular project in which many people might participate. Democracy as a universal ideal was thus born of a devolution of the 'divine corporation' to many self-sustaining *universitates*, the source of modern constitutional polities (Schneewind 1984).

Indeed, rather than reflecting any divine deficiency, this process of ontological devolution reflected an 'optimisation' strategy, whereby God's will is realised by multiple means, the most efficient of which may be indirect, say, as executed by humans acting out of their own accord, which would allow us to be divine creatures and autonomous agents at once. 'Principal-agent', an expression that political scientists and economists nowadays use for the relationship between, say, the people and their elected representatives or a firm's shareholders and its board of directors, captures this newfound sense of God's reliance on humans as his 'agents', the paradigm case of which may be Jesus.

This entire way of thinking about divine causality in nature was indicative of the controversial branch of early modern theology, *theodicy*, which tried to infer the principles of divine justice from a world that is supposedly 'the best possible' yet admittedly full of imperfection and even room for improvement. Theodicy attempted to exploit the obverse of the Scotist thesis that human virtues are diminished versions of divine ones: namely, that God can be understood as a rational calculator with superhuman powers. In that case, crucial to the collective instruction of humanity is the reverse-engineering of nature to learn of its divinely inspired 'functions', so that we are in a position to devise means to provide for nature's ends more efficiently. This mentality, which informed Linnaeus' original classification of species, nowadays goes under the secular name of 'systems-theoretic', where it has remained a powerful heuristic not only in biology but also across the socio-technical sciences, including engineering, cybernetics, economics and of course structural-functionalist sociology (Heims 1991).

In this context, the criteria used for incorporation into such a 'system' – say, via the passing of an examination, the pledging of an oath, the payment of a fee – served as a constant reminder of the sense in which the social order is an open projection: A common past can be extended into a variety of futures, depending on exactly who is selected to reproduce the *universitas*. Unlike a royal dynasty or a caste system, conjugal relations need not provide the default mode of succession: Identity is literally always under construction and potentially available to anyone who can pass the membership criteria, regardless of origins. This idea acquired great force and generality in the modern era through *social contract theory*, an attempt to bridge sacred and secular history in a philosophical discourse inspired by the various mutually binding agreements, or 'covenants', through which God ensured the future of the Chosen People in the Old Testament. In the first instance, it implied a substantial overlap in the divine and the human intellects – certainly enough to allow for the sophisticated transactions described in the Bible that still intrigue game theorists (Brams 2002). But equally, it suggested that while any human project may be of indefinite duration, its continuation requires periodic re-dedication, as exemplified by the dialectic of elections and revolutions that defines the history of modern politics. That dialectic speaks to the centrality of what Max Weber would have recognised a hundred years ago as *decisionism,* the modern remnant of the Scotist idea that in matters of ethics and politics humans partake of divine creativity by the sheer capacity to discount the past and declare one's decision to be the source of all that follows. As briefly discussed in the next section, the philosophy of science has its own version of decisionism, also of the same vintage, called *conventionalism*.

4 Renaturalisation as dehumanisation: The long march back to Scotland

Karl Marx exemplified the bipolar disorder that I have identified with the concept of humanity. On the one hand, his projection of a planned economy in the post-capitalist world would seem to count him as an extreme decisionist. On the other hand, the heightened state of rational will in this Communist utopia marks a radical break with the normal run of human affairs, which he famously characterised as a matter of people making their own history but not as they choose. By this Marx did not mean to suggest that someone (or something) else made it for them, though that might have been a reasonable conclusion to draw from the philosophical historiography of Hegel or any of his precursors in the Enlightenment or those who professed theodicy. Rather, Marx was pointing to a subtle distinction that remains inscribed in the asymmetrical attitudes towards decisions in ethics and epistemology, at least in analytic philosophy: *We decide what to do but not what to believe* (Williams 1973).

The overwhelming significance of unintended consequences in history – that the world does not turn out as we plan – underwrites this asymmetry. However, the asymmetry is by no means absolute, and in fact is really only obvious when the relevant beliefs are clearly evidence-led (Fuller 2003: chap. 11). In contrast, a strong sense of epistemic decision-making is integral to both Protestant theology and positivist accounts of scientific theory choice. Indeed, a broad church of philosophers over the past hundred years, from Nietzsche and James to Carnap and Popper, have asserted the efficacy of decisions by stressing our world-making capacities, which amount to our willingness to follow through on the basis of a set of self-legislated principles, even in the face of some, if not all, empirical obstacles.

Decisionism in this broad sense keeps with the spirit of the two founding figures of modern ethics, Kant and Bentham, who cast their own theories as being about *legislation* – in the former case based on the reasonableness of allowing everyone to take the same decision, in the latter case based on overall consequences for those affected by one's decision. After Schneewind (1997), it is now accepted that this signature feature of modern ethics – whereby one adopts the standpoint of the rule-maker, as opposed to the rule-follower or, still more passively, a being to which rules are applied – is an outcome of the secularisation of theodicy, whereby the sense of cosmic justice pursued by the deity in Malebranche and Leibniz was downsized during the Enlightenment into, respectively, deontological and consequentialist normative ideals of human action, again with Priestley playing a pivotal role in the latter case (Nadler 2008). What then began to slip away, already in the mid-19th century, was the autonomy of the human decision-maker, a point to which the Foucaultian historiography of the social sciences has been especially sensitive.

On the one hand, the social sciences certainly appeared to protect 'all and only' humans as a distinct domain of inquiry and concern. Yet, on the other, as we saw in the case of Marx, social science explanations tended towards the naturalistic, whereby humans appeared more as objects than subjects of history. Indeed, this dose of naturalism – what social research textbooks nowadays operationalise as the 'analyst's perspective' – was probably necessary for social scientists to make a clean break from legal theorists and purveyors of commonsense epistemology. Thus, by the end of the 19th century, the combined efforts of historical linguistics and empirical anthropology had consigned the idea of an original social contract to a category of myth already inhabited by biblical covenants. Nietzsche's 'genealogy of morals', which treats ancient texts as symptomatic of thoughts and feelings continuous with our animal nature, merely vulgarised what was quickly becoming the dominant trend in social sciences, especially once Darwin's *Origin of Species* and *Descent of Man* entered the scene. The fact that such founders of academic sociology in Germany – and younger contemporaries of Nietzsche – as Ferdinand Toennies and Max Weber continued to stress the contractarian nature of modern society should be understood as an attempt to swim against this current that was driving towards a re-naturalisation of the human (Proctor 1991: chap. 8). The sociologists proved more-or-less successful for most of the 20th century, as the ultimate *universitas*, the nation-state, remained the standard-bearer for 'society'. But there is little doubt that their force is dissipating in the face of 'the postmodern condition' (e.g. Urry 2000).

Twentieth century philosophy charted the path to re-naturalisation somewhat differently but largely to the same effect. Some combination of Kant and Bentham, especially John Stuart Mill's rule utilitarianism, became quite influential in disciplines with a strong normative top-down approach to society such as constitutional law, public administration and welfare economics. Rawls (1971) was arguably the last great philosophical moment in that tradition. However, in the aftermath of the First World War, theories of 'mass society' and 'complex democracy' cast increasing doubt on the general philosophical ability to harmonise clashing value-orientations within a single ethical system. Logical positivism's response to this problem set the framework for what followed – namely, to treat ethics as a purely formal discipline concerned with moral language (i.e. 'metaethics'), the content of which is determined by the ideologies of the day. Alasdair MacIntyre (1981) dealt a lethal blow to this approach by arguing that ethics becomes irrational if it is not rooted in the life circumstances of those who are called to live by its principles. He embedded this point in a familiar historical account of modernity as the disenchantment of the world but then gave it a distinctly negative spin. Whereas modernists invoked Kant and Bentham as preserving the best of Christianity without the dubious theological scaffolding, MacIntyre argued that the modernists only succeeded in mouthing rules and principles that in practice did little to constrain people's actual behaviour; hence, the ease with which modernity had slipped into relativism and nihilism.

However, MacIntyre failed to see that the formal character of modernist ethical horizons might reflect the projected experience of creatures who aspire to divinity, and hence to feel *fully* responsible for their actions. Indeed, Kant's and Bentham's universalism bears the clear marks of

Christian origins: the decision-maker's detached divine standpoint combined with an *a priori* egalitarian attitude towards all humans, given the shared divine origin of their souls. Thus, Biblical parables that would have us seek God in the most vile and wretched of our fellows are devices to instil a moral sensibility that countermands our 'natural' response to encounters with particular people, which might be revulsion or sheer avoidance. Legal doctrines of 'procedural justice' that instruct officers of the court to undergo heroic abstraction from the specifics of individual lives function in much the same way, a point that Rawls (1971) especially exploited to great effect.

Thus, the fact that Kant and Bentham appeared to presume that people have much greater knowledge *ex ante* and much greater tolerance *ex post* of the consequences of their decisions than is normally expected in everyday life simply shows that they were trying to raise our normative standards, so that humans cannot be confused with mere animals. Unlike MacIntyre's followers in 'virtue ethics', Kant and Bentham regarded our spontaneous moral sensibilities as more raw material than natural norm. Each in his rather different way would have us develop into creatures who can view, say, the suffering of others as symptomatic of a global disorder that needs to be addressed at the level of principle, which is to say, *not* in terms of our spontaneous response to suffering. Thus, we would want to learn as much as possible from the suffering of others to prevent its future occurrence but without necessarily committing ourselves to relieving the suffering of those who have stimulated this change in course of action.

In contrast to this entire line of reasoning, virtue theorists such as Martha Nussbaum (1986) trust our spontaneous moral sensibilities, which they then try to ground in 'human nature'. Unfortunately, they follow MacIntyre's lead in relying on Aristotle, sometimes as filtered through Aquinas, when in fact the biological basis for human nature has shifted to Darwin, with the emphasis now on the 'nature' rather than the 'human' part of human nature. As a result, virtue ethics, perhaps unwittingly, has contributed to the re-naturalisation of the human, with 'virtue' often nowadays functioning as a quaint word for gene expression. A good way to see this point is by considering 'reciprocal altruism', the principle that evolutionary psychology nowadays postulates as the foundation of ethics, both human and animal (Wilson 2004). This principle supposes that organisms will do what they can to maximise the survival of their common genetic material. It may effectively mean that some organisms sacrifice themselves for others that are better positioned to reproduce their shared genes. Reciprocity may cross species boundaries, given the vast overlap in genetic makeup amongst life-forms. And while the logic behind reciprocal altruism is sufficiently abstract to be easily treated as a version of game theory, it is meant to apply to specific ecologies consisting of well-defined sets of interacting organisms of various species.

This very embodied and embedded sense of mutual concern recalls the spontaneous displays of 'benevolence' that Scottish Enlightenment figures like Adam Smith and David Hume took to be the natural ground of moral judgement. But it stops well short of the universalistic aspirations of a Kant or a Bentham, who would have us act in a way that treats everyone equally, regardless of the personal impact that someone's existence might have on us. Here it is worth recalling, as Ernst Mayr (1982) famously put it, the ontology of species underlying evolutionary thought is one of sustainable populations rather than categorical types. Although Mayr's point is normally treated as pertaining to the nature of biological science, it equally applies to the moral horizons informed by this science. Perhaps the only significant difference between today's evolutionary psychology and the Scottish Enlightenment is that the former tends to stress the *limits* of our benevolence – namely, it fails to encompass those members of our own species with whom we are unlikely to have any evolutionarily relevant contact, notwithstanding our species-typological likeness to them; hence, our persistent deafness to calls to end poverty in areas far from our homes (Slovic 2007).

In contrast, if we spontaneously thought of humans in more species-typological terms, such that we regarded each person abstractly as an instance of the same type to which we ourselves belong, then our practice would conform to the universalist ideals of Kant or Bentham. Rather than thinking of ourselves as we most naturally think of others, namely, as victims of their fate (understood as their inherent dispositions), we would see them as we naturally see ourselves, namely, as masters of our fate who are sometimes confounded by circumstances. In social psychology, this asymmetry in how we explain our own and others' behaviour is called the 'fundamental attribution error' (Ross 1977). More generally, the difference between appreciating someone else's plight at a distance and imagining them as virtual versions of ourselves marks a still underestimated distinction between, respectively, *sympathy* and *empathy* (Stueber 2006).

One of Christianity's great gifts to epistemology has been its insistence that humans are not simply pain-avoiding repositories of experience but agents empowered to organise their experience according to their own designs. In short, it is not enough to sympathise with where our fellows are coming from. We must also learn to empathise by imagining that, under other circumstances, they could be the ones judging us. It is worth stressing that such a position flies in the face of a principled empiricist epistemology, which would reduce us to brutes in Hobbes' state of nature who live in constant fear of one another, in large part because they find each other's actions inscrutable. A cynic, perhaps Hobbes himself, might nevertheless infer from our analysis that empathy is all about insuring against revenge, a thought that at least takes the target of one's concern as seriously as they would take themselves. For Hegel's great theological rival, Friedrich Schleiermacher, our curious capacity for empathy, which he called 'divinatory understanding', is the primary form of hermeneutical knowledge. As if to anticipate Chomsky, Schleiermacher's proof that we possess such a capacity is our spontaneous linguistic creativity, the fundamental expression of *logos* (Schnädelbach 1984: 117–18). In short, we can make sense of others because we can constitute them in our own terms. More recently, this capacity has been given a high-tech gloss in terms of our ability to simulate the lives of others as versions of our own (Stueber 2006). In this context, it would be interesting to reconsider the later Wittgenstein's dissolution of the problem of other minds by rendering the mind intrinsically linguistic and language intrinsically public. Wittgenstein might be seen as playing Spinoza to Chomsky's Descartes.

The secular benchmark for suspending the human halfway between the creative deity and the responsive animal was first laid down by the 18th century Neapolitan jurist Giambattista Vico (1668–1744), who postulated two kinds of sciences – one whose methods apply to all of nature as God's creation and the other whose methods apply only to humans and their creations, by virtue of *Homo sapiens'* unique species-being, epitomised by the reflexively creative capacity of language, which enables us to generate self-contained worlds (aka fictions) within God's *logos*-driven creation (aka facts). Vico's dualistic sense of science began to acquire admirers once universities came to be organised along disciplinary lines, driven by professional rather than liberal education, in the second half of the 19th century. The first high watermark for this perspective was the German Neo-Kantian movement, whose leading proponents included Wilhelm Dilthey and Heinrich Rickert, the former distinguishing the 'human' from the 'natural' on ontological grounds and the latter on epistemological grounds, both resulting in a strong sense of 'two cultures' in 20th century academia, the *Geistes*- and the *Naturwissenschaften*, or as the English simply say: 'arts' and 'sciences' (Collins 1998: chap. 12). Smith (2007) offers perhaps the most sophisticated and comprehensive recent defence of Vico's position as the master builder of the 'human sciences' as they exist today, namely, as those social sciences firmly grounded in textual interpretation and immune to the charms of naturalism.

My own position, in keeping with the Scotist imperative, dissents from Vico and his admirers and is a bit closer in spirit to Noam Chomsky's (1966) attempt to revive the reputation of Vico's historic antagonist, René Descartes, as the great champion of humanity's unique reflexive linguistic capacity. Although Vico was a secular jurist with a strong taste for pagan sources, I believe that he should be read as offering new grounds for re-asserting the position of humanity in Roman Catholic cosmology – indeed, of the sort that his contemporaries would have easily recognised. Vico's distinctiveness lay simply in his efforts to accommodate pagan sources to that cosmology: They are not segregated into a parallel interpretive universe but are treated as themselves sacred sources in defining the nature of human creativity. In this respect, Vico is properly seen as the long-lost godfather of hermeneutics and postmodernism more generally. I have stressed the Catholic provenance of Vico's views because his famous definition of the human in terms of that which we create for ourselves is meant to cast human creativity as only analogous – *but not identical* – to divine creativity. The difference between *analogy* and *identity* is crucial here.

Modern champions of the *Geisteswissenschaften* make too much of Vico's attempts to preserve human distinctiveness from nature, only to ignore the ontological distance that Vico effectively placed between ourselves and God. Vico was keen to maintain a distance that not only Protestants but also heretical Catholics like Galileo and Descartes had reduced in the name of Scotist doctrine of univocal predication. In terms relevant to this discussion, to say that 'God creates' and 'I create' is to use 'create' in exactly the same sense, albeit with the understanding that God creates much better and on a grander scale than I do. In this respect, the modern doctrine of 'reductionism' so closely associated with positivistic approaches to scientific knowledge since the late 19th century is a misnomer. Reductionism is usually criticised – if not outright loathed – for its failure to capture the richness of human experience. But this is to get the wrong end of the stick. From, say, the Cartesian standpoint championed by Chomsky, reductionism should be valorised for its capacity to generate entire worlds from primitive elements and first principles in the human imagination, a micro-version of the divine *logos*.

In contrast, Vico held to a radically bifurcated sense of human cognition. On the one hand, he believed that we could know with certainty in human affairs, even from the distant past, because we could literally rethink the thoughts of fellow human beings. But on the other hand, Vico denied that we could know with certainty in natural affairs because we cannot literally rethink the thoughts of nature's creator, God. Descartes' epistemic reliance on deduction from first principles was meant to do precisely what Vico had deemed impossible on familiar Catholic grounds of arrogating for humanity that which belongs only to God. Indeed, the signature Cartesian claim, *cogito ergo sum,* arguably reproduced God's original self-identification through *logos*. Vico followed the Catholic line that such a move constituted blasphemy. Similarly, Vico would have no truck with the Baconian proposal that nature and Bible are *literally* alternative books for revealing the divine plan, let alone Galileo's claim that mathematics is *literally* the language of nature. Were Vico with us today, he would reject outright the claims made by both Ultra-Darwinists like Richard Dawkins (1999) and intelligent design theorists like Stephen Meyer (2009) that the genetic code is *literally* the language of life.

From a Cartesian standpoint, Vico lacked the idea (or belief) that the human and divine minds, however different in power, overlap sufficiently to enable us, at least in principle, to acquire a systematic understanding of the universe – of the sort associated with modern physics – without ever expecting to experience the universe in its entirety. Descartes' cleverest follower, Nicolas Malebranche, had characterised human participation in the divine mind as our 'vision in God' (Pyle 2003: chap. 3). Inspired by the radical Franciscan William of Ockham, Malebranche argued that we know each other through knowing the same God, in whose image and likeness we are all created. That is to say, we do not have direct access to each other's minds but only indirect access via 'our vision in God', the ultimate ground of our sense of species unity. In philosophy, this doctrine is called *occasionalism*. Its name is meant to suggest that were it not for this common spiritual descent, we could not explain an obvious fact about our knowledge of human beings: We know a lot about people, past and present, without actually having lived lives like theirs and, in most scholarly contexts, without even having made their acquaintance. Leibniz dubbed this cognitive capacity 'intellectual intuition', and philosophers after Kant, innocent of its theological baggage, have been calling it '*a priori* knowledge'. It is also the spirit in which Chomsky (1966) proposed the idea of a 'universal grammar' out of which particular natural languages are constructed. Even in its most demystified form, as in Henri Poincaré's conventionalism, whereby physicists are allowed free rein in their choice of geometry, the God-like capacity to construct worlds *ex nihilo* from first principles has persisted. The latest incarnation of this impulse is the ascendancy of computer simulations as the main platform for the conduct of inquiry across the sciences (Horgan 1996).

Readers may find this line of thought disorienting for two reasons. One reason is the very idea that the human and divine minds might 'partially overlap'. It may be useful to think of this claim as the idealist analogue of the materialist claim that we share virtually all of our genome with the higher apes: In each case, both the vast identity and the marginal difference are meant to carry significance in defining who we are. Consider the most ambitious characterisation of the physicist, 'Laplace's demon', who given the laws of nature and the precise location and momentum of all matter at one point in time could calculate the entire history of the universe. Yet, this reproduction of divine omniscience would not necessarily be matched by divine omnipotence, namely, the ability to bring about the history that the demon could so precisely calculate. The other source of disorientation is the re-positioning of Vico as an intellectual conservative. For Catholics, a great virtue of Aristotle – whom Vico closely follows here – was his recognition of incommensurable modes of being, each with its own characteristic experience. Aristotle's was an ontological, but not an epistemological, relativism: There are multiple modes of being, each with its own specific mode of access. To wish to resolve this multiplicity into one ultimate sense of being with a unique mode of access would be to venture blasphemously into the mind of God.

The various de-humanising horrors conjured up by talk of 'reductionism' are secular descendants of this attitude, the historic antidote to which has consisted in the various 'knowledge interests' and 'value realms' favoured by the Neo-Kantians and phenomenologists over the past century. But this post-Catholic line of thought fails to credit reductionism with the cognitive aspiration for humans to adopt the divine standpoint by considering how the smallest number of principles might generate the widest range of phenomena. What appears complex to us would probably appear simple to God, the origin of all that is and can be. This arresting thought originally lay behind Ockham's Razor, a radicalisation of the Scotist imperative that animated the Protestant imagination during the Scientific Revolution (Harrison 1998; cf. Fuller 2007b: chap. 3).

From this 'Protestant' route to the human sciences that bypasses Vico, how does one justify the distinctive methods of the human sciences, as epitomised by the phrase 'interpretive understanding'? In the human sciences' heyday, say, in the works of Dilthey or Max Weber, these methods were grounded in the biological unity of the human species, which implied that all humans, regardless of their many cultural differences, shared certain fundamental attitudes towards the world, simply by virtue of being human. In Weimar Germany, this assumption came to be elaborated as 'philosophical anthropology', which sometimes overlapped with a strictly biological discipline like ethology (Smith 2007: 46–9). Yet, for the most part, the biological side of this presumed species unity has remained 'black-boxed', even in the quite recent past, say, Giddens (1976), which takes for granted that the biological substratum of all humans is the same at least for historical time, the period that directly concerns social scientists. What is striking about this trajectory is the persistent belief in a strong sense of identity amongst humans, even in the face of mounting empirical evidence from behavioural genetics and neuroscience that our biological arrangements are just as plastic as our social ones (Turner 2002).

5 Recapitulating humanity's bipolar disorder: Towards a theory of mendicant modernities

In the preceding two sections, we have recounted two deep traditions in the history of humanity. The dominant tradition is based on the authoritative Dominican version of Catholicism, passes through Vico, the Scottish Enlightenment, the various ontological pluralists of the modern era, including the Neo-Thomists and the Neo-Kantians, and survives today among those who continue to oppose any rapprochement between the natural and human sciences. The other, more subversive tradition is of Franciscan origin, passes through the Protestant Reformation, Descartes and the rationalist philosophical tradition, and the various 'unity of science' movements from idealism to positivism. The first tradition treats the human as an animal gifted in its adaptations to the environment, the second as a virtual deity aspiring to a universal status that transcends its earthly moorings. In the wake of the French Revolution, the two traditions directly confronted each other's claims on the future of humanity in the persons of Johann Gottfried Herder and Johann Gottlieb von Fichte, respectively, who at the time were competing to claim Kant's mantle (cf. Smith 2007: 40). To put the choice in most general terms: Are we creatures forever materially based in the natural world, however high our status in that world may be? Or are we precisely the part of nature capable of transcending its material basis in order to achieve a higher spiritual unity?

In Herder's and Fichte's day, the policy difference was between focussing on long-standing blood ties or explicitly constituted states. The dichotomy is nowadays most vividly exemplified in the opposing pulls that humanists face from, on the one hand, animal rights activists who would have us reconnect not only with our ancestors but with nature more generally and, on the other, digital media activists who would extend the reach of artifice beyond a written constitution to a computer programme. In section three, I traced this difference in worldview back to the staff of the early medieval universities by the Dominican and the Franciscan friars, the lasting significance of which I shall explore below.

The Dominicans and the Franciscans – or Blackfriars and Greyfriars, to use their English names – are counted as 'mendicant' orders of the Roman Catholic Church, which means that their wealth came not from the ownership of property but from good works, which in turn invited charitable contributions. In terms of economics, this constituted a revolution in the concept of begging, as one's worthiness for charity is demonstrated by the proven ability to do much with little (Langholm 1998). (It is easy to see how this could lead to an argument for 'economies of scale': with more, the mendicants would even be more efficient, given their imperative to minimise material dependence.) When these orders were founded in the early 13th century, they sent a very powerful message about the need for those pursuing a life in Christ to be detached from secular power, such that the fate of the Church itself does not hang on the outcomes of particular wars or power struggles.

This point had special relevance in light of the so-called Investiture Controversy that had been already raging for nearly two centuries, in which the Christian clergy and the secular nobility were making claims against each other's political legitimacy and entitlements throughout Europe. The rise of the mendicants reflected a sensibility that truly came into its own with the Protestant Reformation – namely, a need to recover an original sense of Christian mission whereby standard-bearers of the faith conduct their own lives in an exemplary fashion. It was just this orientation to the world that qualified the Dominicans and Franciscans to teach in universities, whose original 'pay-it-forward' political economy (i.e. alumni donations) was inspired by the example of mendicant living.

While the pursuit of knowledge for its own sake joined the Dominicans and the Franciscans in common cause, they divided over the nature of the good that others – especially students – received from these activities. This division has left a lasting impression in the history of Western inquiry. It was most clearly marked in the academic accreditation associated with the two orders. The Dominicans gravitated towards Doctors in the Professions (especially theology), and the Franciscans towards Masters of the

Arts (especially logic). A doctorate certified that the candidate had established dominion over some intellectual terrain, such that his knowledge 'added value' by providing a deeper or more comprehensive understanding. In contrast, a master's degree demonstrated that the candidate had developed skills that 'saved labour' by permitting a more efficient organisation of thought in public. In the Dominican case, academic skill is an irreducible labour that mediates an increasingly complex society; in the Franciscan case, academic skill reduces labour, thereby removing the social and material barriers to achieving self-legislated goals that remain the source of human misery.

The image of a 'heaven on earth' that the two orders suggested were strikingly different. On the one hand, the Dominicans imagined a multifaceted harmonious world that is captured by the politics of 'subsidiarisation', whereby authority is devolved to administrative units where expertise can be most convincingly and reliably exercised. On the other hand, the Franciscans envisaged a streamlined world that maximised the efficiency with which individuals could pursue their life-plans by providing them with the mental nimbleness needed to find less burdensome means to the same ends, a strategy that we nowadays associate with 'liberalisation'. The Dominican sees the social bond maintained and extended by our recognition of the need for the service of others, the Franciscan by our recognition of the need to serve others.

Pathologies are also associated with the two visions, whose salience increases in the modern era. The greatest Dominican, Thomas Aquinas, troubled over the onset of *acedia* in monks who perform rituals that may indeed benefit others but without investing the relevant spirit. This is the medieval version of today's pejorative account of the bureaucrat as a cynic simply going through the motions. On the Franciscan side, there is the onset of frustration that comes from 'relative deprivation', that is, the current generation's sense of failure to improve as much as the previous generation had. This frustration is potentially the source of violence in a never-ending quest to get more from less, the omega point of which would reach beyond simply explaining or achieving the most with the least effort but would recover God's capacity to create everything out of nothing.

6 The fugitive essence of 'The Human': Towards Humanity 2.0?

I have been tracing the very idea of a domain concerned with 'all and only' humans – the twin root of sociology and socialism in Comte and

Marx and, more generally, social science and democracy – to a specific dynamic conception of *Homo sapiens*. It extends from theological innovations introduced by John Duns Scotus that collectively licensed direct comparisons between human and divine being. Although generally unappreciated by secular philosophers, this genealogy has not escaped the notice of the recent 'radical orthodoxy' within Anglican Christian theology, which treats the Scotist critique of Aquinas as the founding moment of modernity, a version of Original Sin in secular time (Milbank 1990). I largely agree – but minus the negative spin! Scotus diagnosed our difficulty in comprehending God's infinitude in terms of our failure to see how all the virtues of mind and body – which in everyday life are invested in quite different acts, people, etc. – could be equally and ultimately instantiated in God. For example, morally upright people are not necessarily either the most knowledgeable or the most powerful, let alone both. (Indeed, as we shall see in the next chapter, the emerging ideology of 'ableism' outright justifies the selective extension of specific socially desirable traits – say, through biochemical enhancements – independent of possible side effects.) Nevertheless, we might strive to be in ourselves and encourage in others the concentration of these virtues. This idea became the source of the *asymptotic imagination*, according to which we get closer to, if not outright merge, with the deity by making progress on this project, even if that demands a radical reconfiguration of not only ourselves and social arrangements but also our very material nature, including perhaps the 'transhuman' futures countenanced in this section.

In this context, Jesus has exemplified how all the virtues might be temporarily consolidated in a single member of *Homo sapiens*. This was certainly how Joseph Priestley interpreted the life of Jesus, and in the 18th and 19th centuries Newton was sometimes portrayed as the 'second coming' of Jesus, given his own remarkable powers of synthesis, which once again should be understood in the Franciscan sense as a capacity to explain the most by the least (Fara 2002). More mainstream Christians have tended to dismiss such 'Unitarian' interpretations for their apparent denial of the divinity of Jesus. But it would be more correct to say that Unitarians merely deny the *uniqueness* of his divinity. In other words, Jesus enabled others to lead similar lives in their own times, which is something quite different from treating his words as dictations to subordinates. Of course, that still leaves open the equally – if not more – heretical suggestion that Jesus did not quite manage to embody all the divine virtues in his own being, thereby motivating the persistent drive to self-improvement in secular time, a guiding theme of the so-called Radical Enlightenment (Israel 2001).

Here it is worth recalling the common root of scientific theorising and Jesus' 'transfiguration' to Christ – that is, his apotheosis – in the Greek *theoria*, the source of the modern word 'theory'. Jesus became Christ when he acquired the capacity to see everything from God's point-of-view, the so-called 'view from nowhere' first raised in section three. It was then that Jesus caught at least a glimpse of the overall nature of creation, including his own purpose within it. (The relevant Gospel passages are Matthew 17:1–9, Mark 9:2-8, Luke 9:28–36). Of the main Christian traditions, the Eastern Orthodoxy has adhered most strongly to a strong sense of *theoria,* such that humans might become divine at some point in their lives. The history of modern science, given its strong Christian roots, should be understood in this vein, even though the Orthodox Church itself appears to have played a surprisingly minimal role. While this may simply mean that a position that inspires as heresy stifles as dogma, nevertheless the most telling expression of the spirit of *theosis* in our own times comes from the Ukrainian Orthodox Theodosius Dobzhansky, whose dual training in natural history and experimental genetics ideally placed him to be the principal architect of the Neo-Darwinian synthesis in the mid-20th century (Dobzhansky 1937). In a late work, *The Biology of Ultimate Concern,* Dobzhansky (1967) championed the maverick Jesuit palaeontologist Pierre Teilhard de Chardin – about whom more in Chapter 4 – as providing a broadly Christian justification for humans steering the course of evolution through ethically regulated genetic experimentation. Indeed, another of Teilhard's champions, Julian Huxley, coined 'transhumanism' to capture just this perspective, which we shall explore below. All of the above would contribute to a revisionist history of eugenics that is likely to be written later in the 21st century.

However, generally speaking, the 19th and 20th centuries were driven by an increasingly heightened sense of *universitas* – a more collective if not outright corporeal sense of asymptotically convergent virtues. The theological roots reach back to John Calvin's interpretation of the Christian sacrament of the Eucharist, whereby Christian communicants are said to partake of Christ by ingesting bread and wine. Calvin took literally the idea that Christ is embodied in any community assembled in His name for such a purpose. Two centuries later, that heretical native of Calvin's Geneva, Jean-Jacques Rousseau, secularised Calvin's reading of the Eucharist as the 'general will', thereby blurring the individual-collective distinction altogether by defining a normatively acceptable person – what after the French Revolution was called a 'citizen' – in terms of one's willingness to represent the social whole. This move launched an era of 'ideological' politics, whereby people were routinely expected to stand for ideas, even if it meant dying for them in the process. We call the extreme end of this development 'totalitarianism'.

Nevertheless, the search for appropriate vehicles of meaning remains. Indeed, the social scientific residue of Calvin's move may be found in Emile Durkheim's (2001) influential view that religious communities routinely perform a kind of philosophical magic whereby epistemic states are converted into ontic ones. In other words, belief in God is concretised as membership in a divinely inspired community. In Fuller (2007a: chap. 3), I describe this as a translation of a *many-one* into a *part-whole* relationship: in this context, instead of many people already believing in the same deity, each believer increases the likelihood that all will realise in their midst a currently only dimly understood God, resulting in a 'heaven on earth'. They will collectively become, so to speak, the new body of Christ. A related but less materialistic transformation is presumed in routine philosophical realist discussions of the search for truth, which presuppose a final resting place (albeit in the indefinite future), in which the inquirer comes to 'know' in the sense of 'embody' – or 'participate in' (as a Platonist might say) – the ultimate theory of reality.

Today, nearly five centuries after Calvin and over two centuries after Rousseau, we have begun to take the asymptotic imagination one step further by seeking a divine convergence of virtues that challenges humanity's biological integrity through the removal of the organism-environment distinction. This move is epitomised by concepts ranging from 'extended phenotype' (Dawkins 1982) to 'cyborganisation' (Mazlish 1993), depending on whether one stresses the general evolutionary or more specifically human sides of this process. In either case, to recall Marshall McLuhan's (1964) classic definition of 'media', this project is dedicated to 'the extensions of man *[sic]*' (cf. Fuller 2007a: chap. 6), or what, in more religious terms, would be called 'self-transcendence'. As these successive reconfigurations of mind and matter have made clear, the Scotist doctrine left open a profound question: *Does our capacity to come closer to God amount to second-guessing him or, to coin a phrase, 'second-powering' him?*

The two possibilities can be distinguished in terms of what we as creatures *in imago dei* are meant to reproduce of the deity. Here it is worth noting that the classical logical expression of natural laws as universalised conditional statements (i.e. 'For all x, If x is A, then x is B') is ambiguous between the 'second-guessing' and 'second-powering' senses of 'coming closer to God'. An epistemic reading would construe natural laws in 'second guessing' terms as efficient procedures for calculating all the states of the physical universe, given initial conditions. An ontological reading would regard natural laws in 'second powering' terms as the set of parameters within which God brought about those states but could have equally brought about other states (i.e. had 'x' *not* been 'A') – and might do so in the future, perhaps with our help or by what are commonly called 'miracles'. But to appreciate the stakes, the matter is best considered in metaphysical terms.

On the one hand, to second-guess God would be to figure out his plan and manifest it as best we can. A secular residue of this view is the idea that the ultimate scientific theory should provide a true and complete representation of all reality. This is what philosophers call 'convergent scientific realism', a position that perhaps originated with Priestley but is most closely associated with the pragmatist Charles Sanders Peirce, an avid fan of Duns Scotus (Laudan 1981b: chap. 14). It seems to imply that a point will come when the work of God (and humanity) is done – in other words, that there is an 'end of science' (Rescher 1984). On the other hand, to second-power God would be to reproduce the sphere of possibilities from which the deity selected a particular plan to realise, so as to maintain the deity's creative freedom. Among the secular residues of this view is the frequent association of technological progress with an increasing ability to keep options open, often by reversing or mitigating the bad effects of earlier choices. In this context, Popper turns out to be God's best friend.

However, as social science has drifted from its theological moorings, the terms of democracy are increasingly up for renegotiation. On the one hand, normative categories traditionally confined to humans, especially 'rights', are being extended to animals and even machines. On the other hand, there are increasing attempts to withhold or attenuate the application of these categories to the disabled (including the virtually disabled, via antenatal screening), the elderly, the chronically ill or simply unwanted or unproductive humans. This characteristically postmodern 'open borders' approach to life is often presented in a positive light as a triumph for anti-essentialism. However, essentialism about 'the human' has arguably always been about using ontology to stiffen moral and political resolve. It is difficult, if not impossible, to define a system of complementary rights and duties that might constitute 'human dignity' without a clear sense of the range of beings eligible for the title 'human'.

From this standpoint, the confidence with which self-styled 'critical' theorists continue to claim to be able to demarcate the human from the non-human rings hollow. For example, both Jürgen Habermas and Noam Chomsky fall back on language as the defining human trait. But Habermas (1981) does little more than democratise Aristotle's account of the human as *zoon politikon* by extending the intersubjectivity displayed in the public sphere to cover all of social life. In effect, and perhaps in keeping with modern sensibilities, he redefines much of the private as public, which then enables him to include all of *homo sapiens* as properly 'human'. As for Chomsky's appeal to language, it is more sophisticated, based on its endlessly creative character – at least when understood at the syntactic level. This last qualification is important because, over the years, Chomsky has increasingly restricted language's creativity to the sheer combinatorial powers of its grammatical units. Conceding recent research in evolutionary psychology, he has even felt the need to acknowledge that many higher apes are capable of complex forms of signification, though *Homo sapiens* remains unique in its ability to combine signs and build upon them, especially in a self-reflective fashion – at least until evolutionary psychology shows otherwise (Hauser et al. 2002).

Even if Foucault overstated the case that humanity is receding as a salient object on the metaphysical horizon, it would appear that humanity's nature is forever fugitive. Any proposed 'mark of the human' has been the target of simulation (usually by machines) and extension (usually to animals), the overall effect of which has been to cast doubt on the original intuitive applications of 'the human' to members of *Homo sapiens*. Just as computers potentially outstrip us in the capacity for reasoning and information processing, animals increasingly appear more complex and adaptive to nature – and in relation to both cases particular humans start to look deficient. Thus, in 2008, *New Scientist* magazine declared that animals have been found to possess at least six traits previously regarded as unique to humans (Douglas 2008):

1. *Culture* – that is, different lifestyles have been found within the same animal species.
2. *Mind reading* – that is, at least primates can sufficiently recognise each other's mental states to engage in deception.
3. *Tool use* – that is, animals not only make tools but in ways that resemble some of the complexity of human tool-making.
4. *Morality* – that is, animals not only act well towards each other and even members of other species but also can reliably judge others' morals.
5. *Emotions* – that is, animals engage in rituals that suggest the capacity for such complex emotions as grief and awe.
6. *Personality* – that is, members of the same species under similar environmental pressures still display varying attitudes towards risk.

I have argued that it may be that our last distinguishing mark as a species is the overriding value that humans have come to place in the very project of science as aiming for the ultimate theory of everything – regardless of its relevance to our survival, which is to say, in open defiance of Darwinian imperatives (Fuller 2010).

In any case, from both a scientific and a political standpoint, what is worth continuing to defend as distinctly 'human'? Perhaps the most ambitious strategy for projecting the predicate 'human' into the 21st century has centred on 'converging technologies' ('CT') – that is, the integration of cutting-edge research in nano-, bio-, info- and cogno-sciences for purposes of extending the power and control of human beings over their own bodies and their environments. CT, presented with interesting variations, is now a part of the long-term science policy agendas of all the major nations but its roots reach back to Francis Bacon's scientific commonwealth and Count Saint-Simon's utopian socialism. I shall consider this development in more detail in the next chapter.

There are at least six variants of CT, each of which may be associated with the sense in which it would have 'the human' projected:

1. *Humanity Transcended:* Julian Huxley's original sense of 'transhumanism', namely, the return of natural selection to its metaphorical roots in artificial selection, such that humans become the engineers of evolution. One might see this as 'reflexive evolution'. Huxley originally conceived it along classically eugenicist lines, whereby the state would be empowered to make strategic interventions to encourage or discourage the frequency of various traits. In cases of doubt as to the exact genetic component of traits, one would remove social barriers so as to allow 'equal opportunity' (a Huxley coinage) for gene expression. Advances in knowledge of genetics allow these interventions to become increasingly upstream (e.g. antenatal screening), which pre-empt traditional forms of social engineering, such that eventually we acquire Plato-like powers to legislate the desirable distribution of traits in a population, which in principle may be quite different from the traits' normal frequency.
2. *Humanity Enhanced:* Most CT science policy statements focus on the prospect of humans acquiring improved versions of their current powers without the more extreme implications of prolonging life indefinitely or upgrading us to a superior species. For example, the use of nanotechnology to eliminate fatty deposits from arteries or clean polluted water is designed simply to raise people's productivity and quality of life, so that they can contribute more efficiently to the economy and the ecology in their normal lifespans. However, the increasing availability of such enhancements may serve (unwittingly)

to shift the standard of 'normal' human performance, so that eventually those who refuse to undergo the relevant enhancements will be classified as 'disabled'.

3. *Humanity Prolonged:* This CT goes beyond enhancing normal life capacities and towards suspending, if not reversing, age-related disintegration and perhaps even death itself (DeGrey 2007). This aspiration brings together theological (both Greek and Christian) preoccupations with immortality and modern medical science's conception of death as the ultimate enemy that needs to be overcome. It is specifically focussed on extending indefinitely human existence in its prime and hence explicitly raises questions about intergenerational fairness, if not the very need for intergenerational replacement or, indeed, sexual reproduction itself.
4. *Humanity Translated*: This CT is arguably a high-tech realisation of theological ideas concerning resurrection, whereby an individual's distinctive features are terminated in one physical form and reproduced in another. It is usually discussed in terms of the uploading of mental life from carbon- to silicon-based vehicles, typically with the implication that the relevant human qualities will be at once prolonged, enhanced and transcended. But it may also include avatars ('second lives') in virtual reality domains, whereby the individual's human existence becomes coextensive with participation in this post-human translation process.
5. *Humanity Incorporated*: The take-off point for this CT is the legal category of artificial persons and corporate personalities, the medieval artifice of the *universitas*, and includes all of its pre-modern, modern and postmodern affiliates: the extended phenotype (Dawkins 1982), the 'supersized mind' (Clark 2008) of 'smart environments' (Norman 1993) and, most radically, 'systems architecture' (Armstrong, R. 2009), whereby human and non-human elements are not only combined but allowed to co-develop into novel unities. All of these proposals share the idea that humanity's distinctiveness comes from our superior organic capacity to make the environment part of ourselves. As Hobbes had already suggested in *Leviathan*, it is an especially materialist take on the *imago dei* doctrine.
6. *Humanity Tested:* Given CT's quite speculative character, the enthusiasm with which large sectors of the scientific and political communities would embrace such an agenda – and not for the first time – arguably speaks to humanity's superior capacity for experimenting with one's own life and the lives of others. Reflecting the likely, perhaps even disastrous, failure of many of the CT experiments involved in

realising any of the previous five projections of humanity, the focus here is on promoting a culture tolerant of risk-taking, say, by a generous social insurance scheme, a supportive environment for reporting and coping with unanticipated outcomes and a strong sense of an overarching long-term collective project. Mainstream scientists have abetted this tendency by educating the public of the risks they already tolerate, compared to which CT-based experiments do not seem so unreasonable (Nutt 2009).

Whatever is made or becomes of these six projected human futures, they are all true to the spirit of humanism, with the emphasis on the *-ism*. This suffix, which came into its own in the 19th century to name the great political ideologies, marks the birth of our current era of virtual reality. The point of virtual reality is to realise the latent potential of the actual world, typically by getting us to see or do things that we probably would not under normal circumstances but could under the right circumstances. Thus, the appeal of, say, 'liberalism' or 'conservatism' lay in proposing policies that extend tendencies already present in society but which require concerted action to be fully realised. 'Humanism', an ideology retroactively christened by Nietzsche's patron Jakob Burckhardt to refer to a movement traceable to the European Renaissance, should be understood as simply a more abstract version of the same phenomenon. Thus, there is more to humanity than simply the normal conduct of human beings. Only those features of our conduct that distinguish us from other animals are worthy of a 'humanist' project. In that respect, 'humanism' is not only a celebration of our situatedness in nature but also an attempt to create the appropriate distance that justifies our species self-regard. The previously enumerated six human futures address precisely that sentiment.

It is worth recalling just how far we have already gone down the path to such 'transhumanist' futures. Here are some examples that are easily taken for granted:

- The channelling of both work and play through digital media, such that time spent in cyberspace increasingly supplements, if not replaces, time spent outside it.
- Computer literacy is now introduced at the primary school level, if not earlier. The symbiotic relationship of the last two generations of humans with computers has led to a greater tolerance and even transferred affection to androids and cyborgs as 'second selves' (Turkle 1984).

- The extension of the law's jurisdiction into 'second life' and other 'virtual realities', such that an English court has settled a divorce case on the basis of a spouse's adulterous avatar (Morris 2008).
- The increasing use, tolerance and demand for brain boosting drugs, silicon chip implants and gene therapy as psychotropic supplements, comparable to caffeine, vitamins, plastic surgery and eyeglasses – which at some point may be seen as equally necessary for a normal existence.
- The ease with which we resort to 'pre-emptive' interventions, be it to prevent wars, crime, unwanted lives (e.g. contraception, abortion and euthanasia) or even aversion to innovation (aka 'anticipatory governance': Barben et al. 2008). It is very much in the utopian spirit of presuming that we do more than simply treat symptoms and effects – we can manipulate purported causes.
- The ease with which we trade off privacy and security for access, as in the emerging phenomenon of 'cloud computing', which promises to make all information available through overlapping providers – that is, multiply accessible but also traceable (Barnatt 2010). It implies a principled acceptance of the 'humanity incorporated' option above.

Finally, in light of these multiple projections of the human future, how should the history of the concept of humanity inform the future of social science? The guiding insight that connects the theological backdrop of the Scientific Revolution and today's convergent technologies agenda is the idea that biology is a branch of technology, that is, the application (by whom?) of physical principles for specific ends. The theological ground for this interpretation is a rather literal understanding that humans have been uniquely created in the image and likeness of the ultimate artificer of nature. This provides *prima facie* license for the reverse engineering and enhancement of our animal natures (Harris 2007). Ontologically speaking, we can be 'creative' just as God can, though perhaps not to such great effect. However, I say '*prima facie* license' because there may also be epistemic, ethical and political grounds on which to regulate, if not curtail, our godlike powers. For example, we might query the degree of risk at which particular individuals or entire populations are placed by biotechnological interventions or, as Habermas (2002) has emphasised, the degree of 'dehumanisation' as conventionally understood that would result, were such interventions to issue in desirable outcomes on a reliable basis.

But even as these normative matters remain unresolved, we need to be clear which objections to the idea of 'Humanity 2.0', so to speak, are merely epistemic and which truly ethical or political. This distinction is often obscured by pre-emptive ontological arguments about the 'limits' of human nature, in which 'limits' may refer either to our ignorance or our scruples: what we *can vs. should* do. A good case in point is the current internal Christian debate concerning the challenge posed by the most sophisticated version of scientific creationism – intelligent design (ID) theory – to biology's Neo-Darwinian orthodoxy (Fuller 2007b, 2008a).

ID is formally indifferent on the hot button issues that have traditionally divided creationists and evolutionists, such as the age of the Earth and the common descent of species. However, ID retains the creationist idea that science can detect the divine signature in nature, based on the organisation of cells and the systemic features of organisms and their ecologies. This signature goes by names like 'irreducible complexity' (Behe 1996) and 'complex specified information' (Dembski 1998). Interestingly, more conventional Christians, including the Thomists who propagate Roman Catholic doctrine, have condemned ID as heresy. This chapter has offered clues to the basis of such a harsh judgement. For even if conventional Christians and ID supporters overlap in their (typically conservative) moral and political beliefs, they find themselves on opposite sides of what I have called 'mendicant modernities': Whereas conventional Christians follow the Dominican line of treating scientific and religious language as operating in different registers (roughly, literal *vs.* figurative). ID clearly descends from the Franciscan side in accepting a univocal account of language.

Thus, the ID supporter holds that to say that 'life is coded information' or 'cells are nano-machines' is to speak literally, not merely figuratively (cf. Kay 2000; Meyer 2009). Indeed, the blog associated with the ID movement, *Uncommon Descent*, features on its masthead a snapshot of a computer simulation of the bacterial flagellum, depicting it as a miniature outboard rotary motor. Whereas Thomists see this as little more than suggestive imagery, ID theorists hypothesise it to be an ideal version of the real thing, on the basis of which bioengineering research programmes might be mounted. Little surprise, then, that the UK's most vocal ID supporter, Andrew McIntosh, is a professor of thermodynamics whose Young Earth Creationist views have not prevented him from being a leading contributor to *biomimetics*, the study of animal and plant design for purposes of fashioning new technologies (Benyus 1997).

From the Thomist standpoint, if you say that cells are literally machines, then you are on a semantic slippery slope to saying that science finishes the work of theology because God turns out to be a super-mechanic whose *modus operandi* we come to fathom by studying the mechanics of nature – and extending and improving it through our technology (Noble 1997). To be sure, God does not disappear from this picture – after all, we are supposedly his apprentices and stewards. But much of the mystery surrounding religion, especially the grounds for legitimising a priestly class, does seem to disappear. That fact alone can be easily interpreted as heresy and even sacrilege, as of course happened in the Scientific Revolution, which also began as a semantic dispute over whether the ways of the deity can be rendered intelligible through ordinary modes of human expression. If it is possible to read the Bible literally – that is, written in a language that each person is cognitively equipped to understand – then the answer would seem to be yes (Funkenstein 1986).

In more contemporary terms, the more we rely on talk and images of machines, information, computers, codes, etc. to make sense of the biological world – and the more they bear interesting empirical and practical fruits – the more reason we have for taking these so-called analogies literally, both in terms of what they say about the created *and* the creator. Only tact conceals this point. For example, to say that machines and codes are not self-sufficient but require a non-mechanical mechanic or code-maker (a point that ID supporters often raise in self-defence) does not halt the process of demystifying the deity that so worries the Thomists and more conventional Christians. On the contrary, it invites deeper study into how the human creative process works as a source of clues to how divine creativity works. An implicit acceptance of this strategy is already written in the history of science, as psychology and, to a lesser extent, economics and sociology emerged from theology and philosophy.

The Thomists want to keep this all in check by policing the way we talk about things so that theology keeps its subject matter from bleeding away into secular disciplines more than it already has. Here the Thomists make common cause with a curious swathe of Protestant theologians who have interpreted the scientism that drove Germany into the catastrophe that was the First World War – and arguably the nations involved in every subsequent international war – as reflecting an ultimate limit in our capacity to know how to do good. From this standpoint, the promotion of science to a hegemonic worldview should be seen as rooted in a hubris that is a lingering residue of Original Sin. Here a 'fundamentalist' Lutheran theologian of the interwar period such as Karl Barth makes common cause not only with the Thomists but also latter-day postmodernists like John Milbank (1990) and Karen Armstrong (2009) in wishing to restrict science to just one more sphere of a religious life rather than as the sphere – *qua* 'natural theology'

– that will redeem Christian theology's promise of enabling us to reunite with God.

There is no doubt that humans are improving their capacity to manipulate and transform the material character of their being. Of course, uncertainties and risks remain, but in the first instance they are about how costs and benefits are distributed in this imperfect process – that is, unless one has a principled objection to humans being changed in certain ways. But principled objections do not require an ontological basis. Here critics need to catch up with the times. Traditional natural law appeals to 'violations of human nature' of the sort frequently invoked by George W. Bush's bioethics panel increasingly lack intellectual currency, given the socially constructed character of humanity and the anti-essentialism of modern biology (cf. Baillie and Casey 2004). Instead, 'principled objections' to Humanity 2.0 should follow the example of resisters of technological innovations, namely, to argue that unregulated innovation is likely to increase already existing inequalities in society. In other words, an explicit policy of redistribution, which in turn addresses both the external and internal boundaries of 'our' society – that is, who *prima facie* counts as human and to which category of human they most relevantly belong – always needs to attend the introduction of any technology capable of radically redefining the human field.

3
A Policy Blueprint for Humanity 2.0: The Converging Technologies Agenda

The subtitle of this chapter refers to the most ambitious global science policy agenda of the early 21st century (Roco and Bainbridge 2002a), which was produced by Mihail Roco and William Sims Bainbridge of the US National Science Foundation (NSF), the former an engineer in charge of nanotechnology research initiatives, the latter a sociologist in charge of the social informatics unit. Their basic idea was to steer the research frontiers of a select group of cutting edge technosciences so that they 'converge' into a single unified science focussed on facilitating our transition to Humanity 2.0. However, the basic idea harks back to the original 19th century positivists Henri de Saint-Simon and Auguste Comte, namely, to channel the leading edge of the sciences into a master administrative science of 'sociology' that will finally rationalise the human condition. The chapter's first four sections are based on my own three-year study of scientists and science policy-makers throughout the world involved in this agenda. I consider the various political aspirations and envisaged implementations, as well as the subtle but significant reworking of the history of science that is needed to legitimise the agenda. After all, not only positivism as a scientific ideology but also its empirical face, biochemical reductionism, has been out of fashion for quite some time now. Yet, the current fixation on 'nanotechnology' – that is, bioengineering at its most minute yet functional level – promises to revive the fortunes of what had been regarded as a modernist delusion. However, as I argue in the penultimate section, down the road lie some tricky normative issues, not least an enhancement-oriented consumer economy in which conventionally normal people would come to think of themselves and each other as 'always already disabled'.

1 Converging technologies as public relations for science

The wave of converging technologies ('CT') policy initiatives may be seen, in the first instance, as an ingenious strategy to revive Cold War science policy. The Cold War was largely fought by proxy in various 'science races', resulting in what may have been a Golden Age for publicly funded science. The prospect of nuclear annihilation – be it to promote, inhibit or simply understand it – was the cynosure of the period's cutting edge science, as it was held to address entire populations equally. However, the post-Cold War political economy of the past 20 years has witnessed a shift in both sensibility and resources away from these necessarily large-scale, nation-defining scientific projects. Simply following the money, the transition has been from physics to biomedicine. There is a surface and a deep reading of this development. On the surface, funding has shifted, at least in relative terms, from public to private control, the sort of change that economists somewhat misleadingly describe as 'from state to market'. (More accurate would be to say that the state has relinquished some significant aspects of its control of the market.) The deep reading is that the locus of funding – both public and private – has shifted from matters that affect us collectively to those that affect us differentially. Nevertheless, CT should be seen as mainly an effort to 're-collectivise' the interest in the biomedical sciences by stressing potential problems that we all continue to face. After an early flirtation with the threat of biological replacing nuclear warfare, CT advocates now stress potential problems that arise from an increasingly old and mobile population, which raises the prospect of new diseases that need to be either pre-empted or treated effectively.

There is an intellectual precedent for this move, though it is rarely if ever cited – namely, the German 'finalisation' movement of philosophers and sociologists influenced by Jürgen Habermas in the 1970s (Schaefer 1984). According to the finalisationists, once sciences have reached a state of Kuhnian 'normal science', whereby they are solving well-defined problems within an agreed theoretical framework, rather than allowing them to solve ever more technical puzzles, the state should turn their attention to interdisciplinary ventures oriented to solving social problems, such as the then recently launched 'war on cancer'. However, the distinctive feature of today's CT agenda is that the sciences and technologies in question – which usually include at least nanotechnology, biotechnology and information technology – have yet to reach the sort of maturity in theory development that is classically associated with a Kuhn-style scientific paradigm. On the contrary, and especially in the robust form promoted in the United States, the idea is very much one of planning for both the production and the consumption of fundamental developments that have yet to occur, thereby blurring the 'basic/applied' research binary that has dominated post-World War II science policy discourse. The term of art 'anticipatory governance' has been introduced to capture this novel sense of science-by-public-relations (Barben et al. 2008), though a clear precedent may be found in Steve Woolgar's (1991) idea of 'configuring the user' of new software by co-opting potential users (and dissenters) in the early design stages so that the finished product hits its target market.

CT's association with public relations should come as no surprise. Most of what passes for 'science communication' research on both sides of the Atlantic reflects its origins as a public relations concern by authoritative scientific bodies, in light of apparent growing scepticism about the value of science. I say '*apparent* growing scepticism' because no generalised scepticism towards science has ever been documented. Rather, surveys regularly find the following threefold phenomenon: (1) general public enthusiasm for the conduct and products of science; (2) general public ignorance of scientific facts and theories; (3) public criticism of specific forms and applications of scientific research that are perceived to undermine the quality of human (and often non-human) life.

That these three things should be interpreted as adding up to a problem in *public relations* indicates the extent to which scientific authorities had underestimated the issue. Here it is worth recalling the original context of the surveys in the late 1980s and early 1990s, namely, the meltdown of the Cold War political economy of science. No longer enjoying ring-fenced public funding, science – especially basic research – was now forced to sell itself in an increasingly competitive marketplace, which has had profound long-term consequences for both academic and popular understandings of science, most notably the shift in the paradigm case of 'cutting edge science' from high-energy physics to biomedical science. This shift has been accompanied by a recalibration of one's default expectations about science. Thus, scientific research is nowadays expected to be distributed across many sites and performed by people with heterogeneous skills and only partially overlapping interests, either or both of which may shade into traditionally non-scientific areas such as governance and commerce.

An interesting Anglo-Continental divide over how the public relations problem is defined comes out clearly when comparing two recent anthologies. On the English side, it is a matter of scientists themselves having to become better communicators (Holliman et al. 2009). On the Continental side, the concern is the emergence of a relatively autonomous sphere of science communication that mediates between scientists and the public (Cheng et al. 2008). But in both cases, the original problem involves removing a 'deficit' in knowledge of some sort. The overall trajectory of science communication research over the past quarter century has shifted responsibility for this deficit from the public to the scientific community itself. This reflects two trends.

The first is an unprecedented rise in science-related media coverage, which has generated considerable public engagement, not least through the Internet, allowing people to explore opposing sources of information as they never could before. Much to the chagrin of scientific authorities, rather than placating the public, this engagement has only served to raise the level and sharpen the focus of critical discourse about science. Second, the science communication agenda has come to be colonised by academic professionals whose legitimacy depends on diagnosing the so-called deficit as a new form of knowledge worthy of study in its own right. The exact identity of this object of inquiry is disputed but, broadly speaking, it is 'The Future'. This is especially clear in the case of self-advertised innovations such as nanotechnology. Here hyperbolic hopes and fears are projected by both the scientists and the public long before there are any hard results to apply for good or evil. These projections are now dutifully registered and studied by researchers on internet-based, user-driven wiki-media. Arguably such a climate of hype is generating superstitions and delusions that would rival those of any millenarian religion. Science communication researchers increasingly speak of 'anticipatory governance' as the sort of activity that transpires in this prospective realm.

Interestingly, the prospect of the media transforming the conduct and content of science itself goes relatively unnoticed. If anything, scientists continue to feign alienation from media processes, regarding journalism as operating with a corrupt epistemology to which, for better or worse, they are forced to adapt. The scientists protest too much. One small but telling example is that scientists are surprised that journalists regard the mere mention of a scientific event – however inadequately reported – as significant science coverage (Cheng et al. 2008: 83). Yet, does not the same principle apply when scientists and their employers rely on bibliometric data to assess research standing? Both cases involve the generation and interpretation of signals, where 'reputation management' is achieved by strategically branded and placed products in large-scale, highly differentiated markets consisting of cognitively limited traders. In this respect, by virtue of similarities in their social structure, 'Big Science' and 'Mass Media' already share certain characteristics that enable them to co-exist much more symbiotically than either would care to admit. However, the point is easily obscured because the scaling up of science and communication has taken relatively independent causal paths.

Even science communication researchers who valorise a populist 'citizen science' as engaged with professional scientists in the 'co-production' of authoritative social knowledge tend to underestimate the distinctly *epistemic* significance of the media. Instead they treat broadcasts of such public engagement events as consensus conferences and citizen juries simply as better or worse amplifiers for the previously repressed forms of 'local knowledge' represented by the citizens who are now allowed to share the spotlight with the scientists and policymakers. Researchers still do not take proper notice of the role that the capacity to stage, re-enact or simulate events in digital media is acquiring as part of the scientific validation procedure, alongside the controlled laboratory experiment and more conventional computer modelling techniques. In this respect, to adapt the old Gil Scott-Heron song, the next scientific revolution may well be televised.

In what is aptly called 'Hollywood knowledge' (Cheng et al. 2008: 165), consultants shuttling between the scientific and cinematic communities do not merely convey information that each needs to know of the other to get their respective points across. Rather, the scientists and the film makers mutually calibrate their goals and standards of achievement. In particular, film makers not only shape the expectations of their viewers but also fuel the scientific imagination itself, as aspects of complex concepts and situations are heightened, sometimes exaggerated, but in any case extended to their logical extreme – very much in the spirit of the best thought experiments. A striking historical example is Fritz Lang's 1929 film, *Frau im Mond* ('Woman on the Moon'), many aspects of which – from the design of the launch pad to the 'countdown' ritual – were adopted by NASA 30 years later, partly under the influence of Wernher von Braun who had seen the film as a youth. More generally, the changing image of DNA – from, say, the simple combinatorial depiction of the molecule on a primitive computer in the 1973 BBC television series, *The Ascent of Man,* to the three-dimensional dynamic machinery one routinely sees today in science films – illustrates how improvements in media representation have helped to re-orient both scientific and lay understanding of what fundamental concepts are supposed to be about.

The very idea that scientific theories might stand or fall by whether they make for, say, a good cinematic experience may sound frivolous at first, but is it so very different from judging the soundness of mathematical models by the elegance of the simulations they produce on a computer screen? In thinking about such proposed seismic shifts in scientific representation, it is always helpful to take the long view. Recall that among the early objections to the use of experiments as the means for resolving scientific disputes was that they involved too much behind-the-scenes staging and editing that resulted in an artefact with no clear correlate in nature that simply served to manipulate the senses. Indeed, in many cases, the experiments could not be strictly replicated but instead counted on the impressed observer coming to see nature through the principles or phenomena allegedly demonstrated by the experiment. Objections of this sort – notoriously lodged by Thomas Hobbes against Robert Boyle – are usefully seen as an early modern version of Plato's ancient critique of poetry and drama as fraudulent imitation of authentic knowledge and sentiment (cf. Shapin and Schaffer 1985). So too we may come to see the current scepticism surrounding 'science-by-television', once people with the appropriate media production skills join the ranks of scientific validators, following in the footsteps of artisans, mechanics and programmers.

2 The transatlantic stakes in the CT agenda

The US and the European Union (EU) have been vying to control the direction taken by CT agenda, but all indications are that the US is winning this struggle, at least at the level of ideology. However, it remains to be seen whether this palpable change in policy discourse results in long-term substantive changes in science and technology itself. What is at stake in their differences? The US strategy aims to leverage short-term practical breakthroughs in *N*anotechnology into a long-term basic research agenda that would enable revolutions in *B*iotechnology, *I*nformation technology and, most ambitiously, *C*ognitive science. This is encapsulated as the 'NBIC' vision of CT. Underwriting this vision is the idea that 'nano' (i.e. a billionth of a metre) is the smallest manipulable level of physical reality that does not incur quantum indeterminacy. Molecular interventions at this so-called 'edge of uncertainty' can be directed to, say, clear the arteries, repair nerves, etc. Seen in their own terms, as developments within chemistry, these interventions are merely incremental improvements. But what matters are the research opportunities these improvements open up in other fields once they are applied.

The sense of 'convergence' in CT here clearly implicates a general history and philosophy of science in which developments in nanotechnology act as a tipping point for revolutionary change across all of science and technology. Indeed, the CT agenda is new only in its explicitness but not its inspiration, which may be traced to the founding policy statement of the Rockefeller Foundation from 1934 that laid the basis for funding on both sides of the Atlantic for what by the 1950s had become the revolution in molecular biology:

> Can man gain an intelligent control of his own power? Can we develop so sound and extensive a genetics that we can hope to breed, in the future, superior men? Can we obtain enough knowledge of physiology and psychobiology of sex so that man can bring this pervasive, highly important, and dangerous aspect under rational control? Can we unravel the tangled problem of the endocrine glands... Can we solve the mysteries of various vitamins...Can we release psychology from its present confusion and ineffectiveness and shape it into a tool which every man can use every day? Can man acquire enough knowledge of his own vital processes so that we can hope to rationalize human behaviour? Can we, in short, create a new science of Man? (Morange 1998: 81).

If we set aside the somewhat dated preoccupation with sex, glands and vitamins, the rhetoric could have come from Roco and Bainbridge's 2002 NSF document. As it turns out, the author of the 1934 statement, Warren Weaver, envisaged the field he coined as 'molecular biology' to be fixated on the phenomena of life at the edge of quantum indeterminacy but still within the range of classical mechanics. Thus, we should come to make very fine-grained positive interventions into organisms without adversely disrupting their systemic functions. This is precisely where the magic of nano-biotechnology is supposed to lie today.

To be sure, the Rockefeller Foundation and the NSF have operated under somewhat different sociological conditions. Weaver was inclined to treat the still novel Heisenberg's Uncertainty Principle as a temporary barrier to human mastery of microphysical reality rather than an insurmountable limit to our understanding of nature. His encouraging the flow of physicists and chemists into biology was designed to demonstrate that point. In contrast, while the NSF document's principal author, Mihail Roco, may harbour similar views, a more pressing policy concern is the decline in employment prospects and, more recently, academic enrolments in physics and chemistry, in light of post-Cold

War shifts in scientific demand – and not only in the US. Science journalists have been especially sensitive to this 're-branding' exercise. Consider this analysis (Ball 2003):

> In March [2003], the Royal Institution (RI) in London hosted a day-long seminar on nanotech called 'Atom by atom', which I personally found useful for hearing a broad cross-section of opinions on what has become known as *nanoethics*. [...] First, the worry was raised that what is qualitatively new about nanotech is that it allows, for the first time, the manipulation of matter at the atomic scale. This may be a common view, and it must force us to ask: how can it be that we live in a society where it is not generally appreciated that this is what chemistry has done in a rational and informed way for the past two centuries and more? How have we let that happen? It is becoming increasingly clear that the debate about the ultimate scope and possibilities of nanotech revolve around questions of basic chemistry [...]. The knowledge vacuum in which much public debate of nanotech is taking place exists because we have little public understanding of chemistry: what it is, what it does, and what it can do.

In short, we may be living in a time when Weaver's ambitions are being revisited to good effect by CT, albeit in the spirit of regaining lost advantage and perhaps even lost collective memory of that advantage, all historic spurs to entrepreneurship (Brenner 1987).

In contrast, the EU strategy discusses CT in more modest terms, allowing for multiple convergences amongst different disciplines. Indeed, it is ultimately less concerned with the future direction of science than on what Joseph Schumpeter meant by 'innovation', that is, the conversion of an invention to a successful market product. The background assumption here is that the scientific community does not provide sufficient incentive to exploit the full social and economic benefit of its new ideas. Under the rubric of CT, the EU proposes incentives to break down cross-disciplinary barriers to enable new ideas to be brought to market more effectively. At the same time, the EU sees itself in a more regulatory role. Where the US initiative calls on both the state and business to reinforce already existing trends in nanotechnology, the EU initiative is much more explicitly about the reorientation of scientists' behaviour from their default patterns to what the 2004 EU report edited by philosopher Alfred Nordmann (2004) called 'shaping the future of human societies'. It was against this ideological backdrop that one of Nordmann's US students, Ashley Shew, attempted to provide the first code of professional conduct for nanotechnologists (Shew 2008).

What might be called the 'dark side' of the idea of convergence consists of research alternatives that are implicitly eliminated – what economists call 'opportunity costs' – as research trajectories are encouraged to come together. Here too we see a difference between the US and EU approaches. There are two general ways of conceptualising this progressive elimination of alternatives: one involving *positive*, and the other *negative*, feedback loops. While there are examples of both types of feedback loops in the interviews and the policy documents, generally speaking, the US CT strategy is given more to positive feedback loops, and the EU CT strategy more to negative feedback loops. In a nutshell the difference is as follows:

- *Positive*: Only certain strands of research provide increasing returns on investment, which in turn attract subsequent resources into those established paths. Policymakers see themselves here as simply adding forward momentum to convergences that, however tentatively, are already taking place (Arthur 1994).
- *Negative*: Research futures are conceptualised here as much more open, which means that policymakers play a greater role in steering researchers in the direction of various desirable convergences that might not otherwise take place, actively discouraging, say, more traditional mono-disciplinary research.

The difference between feedback loops reflects the extent to which CT policymakers see themselves as moving with or against the default patterns of scientific inquiry. In many instances, this difference may turn out to be more of rhetorical emphasis in the formulation of policy statements. However, matters of substance may also be at stake.

CT through positive feedback loops

The US CT stress on positive feedback occurs on two levels: in terms of (1) the strategy used to chart NBIC advances; (2) US responses to those developments. Let us take each in turn.

1. The US government, largely through the initiative of Ron Kostoff at the Office of Naval Research, has invested significantly in 'literature-assisted discovery', which uses bibliometrics to chart rapidly expanding fields in order to anticipate the next stage in a research trajectory,

which oneself or one's competitors may be better positioned to make (Kostoff 2005). The impetus for this investment has been the rapid growth of China's involvement in nanotechnology, making it the world's leader in terms of sheer quantity of published research. However, the quality of the research is still in question, at least as measured by the quality of the journals where that research is published. But that too is improving, as Chinese authors form an expanding portion of those publishing in Western nanotech outlets (Kostoff 2006). The US strategy is to keep constant the goals of CT in terms of 'improving human performance' but remain open-minded about the exact means by which science will serve those goals – that is, by whatever research trajectories happen to bear fruit, which in turn can be used to leverage further basic research. Implied here is a very strong faith in science's capacity to turn up something that will be to humanity's benefit.

2. The US appears willing to let the Chinese strike out in many different nanotech directions, while the US develops 'pipelines' to take maximum advantage of whatever breakthroughs are made. Two pipelines promoted by Roco at the NSF are particularly relevant: (a) The *Integrative Graduate Education and Research Traineeship Program* (IGERT), whereby PhD students are subsidised to work on CT-related projects to counter the department-based allocation of scholarships for doctoral training, perhaps ultimately breaking down the default disciplinary basis for the reproduction of academic knowledge. At a cognitive level, IGERT aims to enable students to think in terms of CT at the outset of their career rather than be forced to synthesise different disciplinary agendas later. A suggested consequence of IGERT is that the next generation of scientists will be more instinctively sensitive to market-driven concerns. Israel and Australia have adapted to the IGERT scheme in contrasting fashion. On the one hand, Israel has built entire universities around the CT agenda, and through the Talpiot scheme provides incentives for younger researchers to get involved in CT. On the other hand, Australia has taken a more nuanced line, seeding some CT-oriented interdisciplinary undergraduate and graduate programmes but typically at lower ranked universities struggling with falling physics and chemistry enrolments, anticipating that students so trained will be best suited for the expanding labour market for lab technicians and research administrators, rather than front-line researchers. (b) The *Industrial Research Initiative* (IRI), whereby US companies develop 'CT platforms', i.e. research capabilities that allow for speedy development of new NBIC-based products. A similar initiative has been launched in Nagoya, Japan. Roco (2002) contrasts

this 'fast but focussed' view of CT's future with that of the more 'science fictional' approach associated with Eric Drexler and Ray Kurzweil (about whom more below). For example, IBM and Intel are investing in CT to find cheaper substitutes for the current electron charge basis of information transmission.

These pipelines are to be facilitated by increased national funding (perhaps with matching corporate sponsorship) for research designed to 'reverse engineer' the brain to enable the more efficient uptake of new knowledge by the appropriate sensori-motor modalities and cognitive faculties. Financial matters aside, the main obstacles to making advances in these areas may be more ethical than technical: i.e. potential so-called enhancement technologies will probably develop faster than public willingness to test and use them. But let us suppose the pipelines proceed as planned. One negative unintended consequence may be major short-term economic dislocation (i.e. unemployment, company closures, investment losses, loss of productivity), as nanotechnology becomes a 'general purpose technology' (GPT) whose innovative and improving cross-sector pervasiveness effectively restructures the entire economy. Such a system realignment occurred in the 1970s and 1980s as information technology became a GPT (Helpman and Trajtenberg 1994). However, at this point the evidence is inconclusive, especially since so much nanotechnology simply extends research in existing fields under a different rubric (Youtie et al. 2008).

CT through negative feedback loops

On the negative feedback side, consider the European Commission communication, 'Nanosciences and Nanotechnologies: An action plan for Europe, 2005–2009' ('NN'), opens with the concern that European scientists are not sufficiently 'entrepreneurial' in the strict Schumpeterian sense of converting inventions to innovations, i.e. bringing their ideas to market. NN goes on to propose various measures to ease the commercialisation of nanotech innovation, including the harmonisation of patent standards and the monitoring and publication of innovation waves. NN also makes a larger and subtler move: It implicitly redefines 'scientific creativity' to mean the sort of mind that sees the commercial potential in new knowledge. Accompanying this definition is a general proposal for reforming science education to bring it closer to a business mentality that blurs the distinction between a university department and a corporate R&D division. While NN clearly aims to advance the CT agenda by counteracting scientists' default tendencies, some quite deep, it is unclear the extent to which these tendencies are simply institutional or more personal.

The original 2002 NSF report has had a demonstrable impact on scholarship, decisively shifting the default meaning of the phrase 'converging technologies'. Roco and Bainbridge (2002b) is the most highly cited version of the report in the academic literature. The various EU responses, starting with Nordmann's 2004 report, have had much less impact, usually only as a critique of the original NSF report. A survey of the phrase in the titles, abstracts and keywords of publications included in the Web of Science and Google Scholar, revealed its pre-2002 occurrence mainly in two contexts. One was in the 'management information systems' and 'knowledge management' literatures, where CT pertained to the integration of information sources as a key to business efficiency in a time when an increasingly dispersed and mobile labour force made it harder for companies to retain the knowledge they had accumulated. The other context was multi-modal educational delivery systems that encouraged 'interactive' and 'distributed' learning regimes centred on student needs and interests. However, after 2002, the use of CT shifted to the scientific project envisaged in the NSF report, though often retaining some of the pre-2002 connotations. Thus, bioinformatics is now often highlighted as a knowledge management strategy for achieving CT, while CT-driven breakthroughs may enable more effective educational delivery systems that reflect and facilitate the brain's capacity to process information.

Lurking beneath differences in formulation, the alternative US and EU versions of CT tap into radically different sensibilities that are somewhat occluded by euphemisms. In the US case, the phrase 'improving human performance' can be sharpened up to refer more explicitly to a project of enhancing individuals by making them – and their offspring – smarter, stronger, etc. This project presumes a sense of biological evolution that might be expedited to the overall benefit of the species by interventions at the level of individual species members. In the EU case, the phrase 'shaping future societies' suggests a more holistic and less invasive approach that focusses on enabling people to live more sustainable lives, where the state or some inter-state authority like the EU is seen as the protector of social equilibrium. In terms of contemporary ecological politics that I shall elaborate below, the US approach is *proactionary* and the EU approach *precautionary*. The term 'proactionary' is a neologism of the transhumanist philosopher Max More (2005), which is meant to be stronger than the Popperian reversibility of piecemeal social engineering because the idea is not merely to reverse a course of action that has already generated negative consequences but to undo those very consequences.

Both the proactionary and precautionary approaches contain ambiguities. In the US case these centre on the meaning of a term like 'improvement' or 'enhancement'. Is one referring here simply to systematically induced changes in, say, genetically controlled behaviour or neural circuitry, regardless of their results? Or does one also wish to imply that these changes are always, or even largely, beneficial? After all, a likely long-term consequence of a US-style improvement policy is an increase in people's willingness to make risky interventions at the genomic or neurophysiological level. But given the complexity of the contexts in which such interventions would play themselves out, their exact efficacy, let alone relative benefit *vis-à-vis* non-intervention, would be difficult to assess. Under the circumstances, an implicit goal of the US approach must be for people to see their bodies as sites of experimentation.

In the EU case, the ambiguities centre on its attitude towards 'marketisation'. On the one hand, the EU clearly wants to remove barriers to the promotion of CT-related innovations that have been erected within but also imposed on academic research. The former refers to the legitimation of inquiry on narrowly disciplinary terms, the latter to legal restrictions on the pursuit of intellectual property rights by public institutions. This is a problem that the US resolved by enacting the Bayh-Dole Act in 1980, which gave the universities the right to commodify their own knowledge work and products and hence the potential to dictate the terms of the knowledge market. Daniel Greenberg (2007), certainly the most venerable and perhaps still the most critical US science policy journalist, has wondered why American universities have not made more of this opportunity to name their price.

India and Germany offer interestingly contrasting adaptations to this development. On the one hand, in India's version of the Bayh-Dole Act, the conferring of intellectual property rights allow universities to increase their corporate autonomy not simply by becoming financially independent of the state but more importantly by laying claim to venture capitalist professors who currently take full advantage of their universities' resources while maintaining exclusive control over their profits. On the other hand, and more in the spirit of NN above, has been Germany's Employee Discovery Law (2002). Formerly German academics were free to collaborate with industry, but afterward academics were treated as employees of the university, which formally owned the intellectual property. General acceptance of this shift in the legal status of the academic from civil servant to entrepreneur has been aided by a massive generational shift, as the '68ers' have made way for academics who have witnessed only increasing neo-liberalisation over the course of their careers.

On the other hand, the EU clearly has a protective attitude towards the public destined to be exposed to the innovations unleashed in such a liberalised economic environment. It would seem then that increased openness to the marketing of innovative products is to be matched by increased monitoring and possibly control of their consumption. This is likely to result in conflicts in the legal system, as both producers and consumers each assert their enhanced sense of 'rights'. I shall suggest below that unlike the US, the EU retains a response mode characteristic of the first crisis of the welfare state as it tries to deal with the second one.

At the level of political economy, the CT agenda may be seen as a 'technological fix' for the second of two fiscal crises of the welfare state that has affected both sides of the Atlantic. The first fiscal crisis occurred in the 1970s, with the increasing tax burden on individuals and businesses to finance wider state coverage of welfare needs. Because this problem was predicted to escalate as more countries reached the standards of living enjoyed by the developed world, calls were made to restrict population growth, via mass contraception and perhaps even some reintroduction of eugenics, especially in the developing world (though 'zero population growth' was portrayed as an ideal in the developed world). What is of interest here is that this technologically-oriented solution diagnosed the problem, in Malthusian fashion, as one of *overconsumption*. However, in retrospect the end of the first fiscal crisis came not from the proposed technological fix but the weakening of welfare state coverage, in the name of 'neo-liberalism'.

The second fiscal crisis of the welfare state, dating from the 1990s, pertains to the anticipated financial burden on the pension system of people living longer after retirement. CT is relevant to this development, as it promises – in both its US and EU guises – a longer period of labour productivity, expanding the economy in general and deferring the need for individuals to draw on pensions. Note that this problem arises in the context of relatively stable, or stabilising, population growth rates. This second fiscal crisis is diagnosed, in Ricardian fashion, as one of *underproduction*. This shift from overconsumption to underproduction is interestingly reflected in the role played by ecological considerations in each: In the former case, nature provides an ultimate irreversible barrier, resulting in a *precautionary* principle; in the latter, nature is a constraint that can be strategically manipulated, resulting in a *proactionary* principle. Indicative of the latter position is the prospect that nano-machines might someday, and perhaps regularly, reverse the effects of industrial pollution in a 'cake and eat it' scenario. This helps to explain the attraction of the CT agenda in the rapidly industrialising economies of China and especially India, where 75% of the inhabitants still lack clean water and sanitation. However, if one regards anthropogenic industrial pollution as an eco-level disease, then nanobot-based solutions may end up creating the equivalent of a drug dependency (Kostoff et al. 2007).

3 Defining 'convergence' in converging technologies: Ontological levelling

For technologies to converge, they must do something more than simply engage in 'synergy' or 'multi-', 'inter-' or even 'transdisciplinarity' (Gibbons et al. 1994). And while the convergence of technologies may produce 'emergent technologies', in the sense of innovations that could not have arisen without the convergence, technologies may also 'emerge' as by-products of the normal development of a single technology. In terms of these nuances, US policy documents are much more explicitly committed to convergence than the EU documents. In the EU context, extended collaboration between two disciplines counts as 'convergence' (Bibel 2004). In particular, BIO + INFO and, more recently, NANO + BIO tend to be targeted as the pairs with the most research and development potential. (By contrast, India's National Knowledge Commission 2006 report defines 'convergence' simply in terms of the involvement of INFO in any novel transdisciplinary research, while in Israel it has referred mainly to INFO + COGNO via linguistics.) Again unlike the US case, there is little talk of forward momentum towards a convergence of many disciplines in the promotion of some overarching goal. Instead the EU model seems to be based on a modified 'finalisationist' model, mentioned at the start of this chapter, which presupposes that disciplines have reached a certain level of maturity that enables them to be steered toward collaboration for socially beneficial purposes (Bibel 2004: General Recommendation 6).

At the most basic level, the idea of converging technologies presupposes that multiple technologies are coming into increasing but also more focussed interaction. The idea stops short of presupposing a specific target but it does contain the idea of an outer limit that somehow shapes the interaction. This point of definition is illustrated in three cases where 'convergence' has a specific meaning in the arts and sciences:

- In art history, a linear 'Euclidean' perspective is defined as a convergence in lines of composition towards a vanishing point on the horizon. The result is to give a sense of closure to a pictorial image that might otherwise appear open-ended and disorienting, as governed by a hyperbolic geometric perspective. This difference has had profound implications in the rise of modern mathematical science as the quest for a unified conception of reality (Heelan 1983; Feyerabend 1999).
- In the philosophy of nature, there is a theory of 'convergent evolution', derived from Jean-Baptiste Lamarck and associated with the heretical Jesuit paleontologist, Pierre Teilhard de Chardin (1961: 238–42). He predicted that, through increased interbreeding and other forms of communicative interaction, human biological differences would be overcome and we would end up turning the earth into a single 'hominised substance'.
- In the philosophy of science, there is a theory of 'convergent scientific realism' associated with the US pragmatist Charles Sanders Peirce. His idea was that through a fallible process of successive approximation, scientists starting with disparate theories eventually arrive at an account of reality that commands the widest possible assent over the widest range of propositions (Laudan 1981b: chap. 14).

As the above examples illustrate, 'convergence' implies that formerly distinct lineages come to lose some, if not all, their differences in a moment of synthesis. This is much stronger than the simple idea that different disciplines share some things in common. For convergence, such commonality must also cause the disciplines to see their interests as more closely aligned, so that they come to orient their patterns of work to each other.

The recent history of the sciences most closely connected with the CT agenda offers some templates for the move to convergence.

- The development of X-ray crystallography in the 1940s first enabled the mass migration of physicists and chemists to biology, eventuating in the revolution in molecular biology associated with the discovery of DNA. The value of this technique was the clear visualisation of phenomena it afforded, most popularly in the double helix structure of DNA. This in turn decisively shifted biology's intellectual centre of gravity from the field to the laboratory, drawing together biology's disciplinary horizons with those of the physical sciences. The physical scientists most attracted by this move also tended to be undeterred by the 'randomness' of nature, be it in the sense of quantum mechanics or genetic variation. They treated life as essentially an engineering project. This migration arguably represents the vestiges of 'biophysics', as inspired by Erwin Schrödinger's famed 1943 Dublin lectures entitled 'What is life?' (Schrödinger 1955; Rasmussen 1997). CT attempts to repeat this movement by enabling people trained in physics and chemistry, fields now subject to declining enrolments and research funding, to migrate to 'nano-bio' fields.

- In the 1950s, a similar development occurred with respect to linguistics, formerly also an archive- and field-based subject based in philology and anthropology. Once a critical mass of data had been gathered on the world's languages, people trained in mathematics and the nascent field of computer science (often under the guise of 'information and communication theory') analysed the sound patterns and grammatical structure of utterances, first in purely statistical terms but later in the attempt to identify 'universal' formal properties. The seminal convergence moment here occurred when Noam Chomsky, one such mathematically trained (and philosophically informed) linguist, turned the tables on his teacher Zellig Harris by arguing that mathematics could go beyond providing an analytic tool to reveal the 'deep structure' of language, the so-called universal grammar that by the late 1960s came to be associated with the still larger convergence of 'cognitive science'. Indeed, cognitive science has been most explicit in its aspiration to convergence amongst the various field-, archive-, lab- and computer-based disciplines associated with the study of thinking, and perhaps even CT more generally, even though its rather theory-driven *modus operandi* does not lend itself to easy integration.
- In the past half-century, computer simulation has become a *lingua franca* for an increasing number of scientific disciplines, enabling the translation and integration of phenomena gathered from disparate sources into a common 'virtual reality' that is projectible and manipulable along several spatial and temporal dimensions. This use of the computer simulation as 'trading zone' for the interaction of different disciplines originated with the Monte Carlo simulations used in the design of the original atomic bombs (Galison 1999; Mirowski 2002: 351–5). Perhaps the most notable site of convergence here has been bioinformatics, whose innovations in

information storage and retrieval allow researchers to pool and share results relating to the testing of various molecular combinations for their biomedically relevant consequences. In this context, genetic information is treated as literally, not metaphorically, digital (e.g. Robbins 1996; cf. Kay 2000).

All of these developments have served to remove traditionally discipline-based barriers to scientific communication. In that respect, they provide for one of the preconditions for convergence, namely, the intensification of researcher interaction. But they also point to a deeper sense of convergence: namely, disciplines are regarded more in discursive than ontological terms. In other words, they are distinguished more by the language they use than the reality they access. Thus, in various cases, the distinction between literal and metaphorical language falls by the wayside: On the one hand, the carbon-based molecular structure of bionic computers enables the solution to problems that have eluded traditional silicon-based computers (Adleman 1994). On the other hand, the structure of DNA itself has been used as the template for the computer architecture (Chang 2003).

Generally reflective of this blurred distinction between the model and the modelled has been the field of *artificial life*, which has shifted its research project over the past ten years from *simulating* to *instantiating* life. The implication here is that carbon-based 'wetware' of flesh-and-blood organisms is no longer regarded as the 'real' or 'natural' form of life that 'software' (i.e. computer programmes) and 'hardware' (i.e. robots) simulate to varying degrees. Rather, life is defined in terms that are completely abstracted from its mode of realisation so that wetware, software and hardware all instantiate 'life' in exactly the same sense (Helmreich 1998; Amos 2006).

The language of 'instantiation' derives from theological discourses of the Christian deity's triune nature, that is, the idea that God is subject to three equally divine manifestations: Father, Son and Holy Spirit. These theological roots go beyond historical curiosity to a general principle of Biblical interpretation that provides a precedent for reducing, if not erasing, the difference between processes, entities and interventions of 'artificial' and 'natural' origin. As we saw in the previous chapter, this principle, associated with what after the 14th century scholastic John Duns Scotus is called the 'univocity of being', takes humanity's creation 'in the image and likeness of God' rather literally, such that human differs from divine creation only in degree not kind: God may be infinitely more powerful than us but he works in largely the same way, i.e. by adhering to the same principles, such that it makes literal sense to speak of humans possibly 'playing God' (Noble 1997). The centrality of this idea to the 17th century Scientific Revolution is very well documented, and helps to explain the Protestant-friendly character of the revolutionaries (Harrison 1998).

When 'life' is treated as an abstract entity subject to multiple instantiations, it is sometimes defined in *functional* terms, such that an artificial entity counts as living if it can pass for a natural life form, as in a Turing Test. However, increasingly the terms in which life is defined are purely *formal*, as in entities that through self-organising means evolve to a certain level of complexity and stability, even if this happens entirely in virtual reality.

A good example of this purely formalist conception of life that played a remarkable role in a legal setting is Avida, a computer programme designed to generate 'digital organisms' (aka computer viruses) according to parameters for self-replication and mutation that approximate those postulated by Darwinian natural selection (Lenski et al. 2003). That after a reasonable number of generations Avida generates stable complex organisms comparable to those in the natural world was offered as evidence for the existence of natural selection in *Kitzmiller vs. Dover Area School District*. The defendants in this US circuit court case had offered intelligent design (ID) as an alternative to Darwinian natural selection, which they regarded as no more than a 'theory' of the origins and maintenance of life on earth.

In this context, it is striking that the judge who ruled against the defendants took at face value the claim that Avida *instantiates* natural selection, thereby obviating the need for alternative theories to be taught (especially given ID's transparently religious inspiration). Thus, even if the exact role of natural selection (*vis-à-vis* other evolutionary mechanisms like random genetic drift and orthogenesis) in the history of natural organisms remains an open question, its general biological validity has been secured by a computer programme that demonstrates the efficacy of natural selection on digital organisms. Perhaps without realising it, the judge had contributed to the CT agenda by granting the same evidentiary status to evolution happening to carbon and silicon-based life forms (Pennock 2005: 91–2).

But the issue of convergence goes beyond accepting different bodies of evidence in support of a common theory. It would be easy to imagine an Avida-like programme interfacing with other programmes responsible for regulating natural organisms to produce a more authentically Darwinian sense of natural selection. I have in mind here the ever-present threat of computer viruses capable of paralysing society's information and communication flows, thereby jeopardising people's livelihoods and even lives. The turn to artificial life invites us to think of this prospect as akin perhaps to cancer, asphyxiation, or simply releasing organic waste from labs and factories into public water supplies and sewage systems. In this respect, the products of computer simulations are not only just as abstract from natural phenomena but also just as real as those of laboratory experiments. Interestingly, post-9/11 national security interests of this rather forward-looking sort initially led DARPA, the US Defense Department's advanced research unit, to support the CT agenda. However, as the global 'war on terror' settled into more conventional modes of warfare and intelligence gathering, Congress stopped DARPA-related CT in 2003 – despite the diligent and imaginative efforts of Ray Kurzweil (2006) to revive CT's military relevance.

Notwithstanding this political shortfall, the potential long-term policy implications of this suggested ontological convergence of the 'natural' and the 'artificial' – both subsumed under a unified conception of 'information' – are enormous. Indeed, the growing legal salience of animal and android rights, especially in light of cyborganisation, makes it increasingly difficult to distinguish where the 'human' ends and the 'non-human' begins. But do they imply that the CT agenda is either 'reductionist' or 'holist'? Some commentators clearly see CT as constituting a revival of the reductionist scientific research programme that would portray all the objects of science as some complex extension of the fundamental particles and forces studied by physics. These commentators tend to stress the particular emphasis that CT, especially in its US guise, places on the nano-level of reality, stressing its drive towards miniaturisation. In that respect, CT appears to be about 'converging downward' to some ultimate constituents of matter. In contrast, support for the holism of the CT agenda rests on its aspirations to create an interdisciplinary or even transdisciplinary science base that addresses questions concerning the enhancement of human performance (US) or welfare (EU) that are not adequately addressed by the individual sciences on the CT agenda. This is, so to speak, a 'converging upward', which is indeed how CT is frequently depicted in the founding policy documents (Schmidt 2007).

However, neither reductionism nor holism adequately captures the distinctiveness of the CT agenda. In particular, it would be a mistake to regard CT as simply a high-tech repetition of the issues classically raised by physical reductionism, in which all of reality is seen as a hierarchy of increasingly complex molecular structures, ranging from subatomic particles to entire ecosystems. Indeed, the *verticalist* imagery of 'top-down' and 'bottom-up' may be itself profoundly misleading as a basis for conceptualising the policy implications of CT. For example, the sorts of hybrid entities generated by processes associated with CT, such as genetic modification, xenotransplantation, computerisation, while generally quite strategic and deliberate (and hence not 'bottom-up' in the traditional sense of 'unintended' and 'emergent') are without any overarching sense of plan that these interventions are meant to serve (and hence not 'top-down' in the traditional sense of 'holistic' and 'preordained'). Thus, even a sophisticated philosophical analyst like George Khushf (2004) stays within the verticalist frame when envisages NBIC as encouraging reciprocal feedback relations between top-down and bottom-up organisations of matter. He misses the strategic character of a theory designed to justify interventions specifically at the nano-level of reality, which is neither the largest nor the smallest units of what is known.

This strategic feature of CT's ontology is characteristic of a trial-and-error 'bioprospecting' mentality that was anticipated nearly two decades ago by Walter Gilbert (1991), Harvard's professor of molecular biology, who was concerned for the intellectual future of his field, as researchers seemed to be content with testing out molecular combinations for their consequences, especially their biomedical uses, but nothing more theoretically interesting. It implies a *horizontalist* imagery, whereby disciplines are linked by common methods – broadly defined as 'modelling techniques' – that in the long run break down disciplinary differences, while reifying the methods as a shared reality. Thus, bioinformatics, originally a tool of molecular biology, becomes the thing of which molecular biology is itself an application.

In this respect, both the US and EU policy documents relating to CT may be seen as providing a focus that tries to reinvent a verticalist perspective to provide an easier basis for governance. Admittedly, the focus in the US and EU documents is defined somewhat differently: 'enhancement of human performance' (US) *vs.* 'improving human welfare' (EU). However, both introduce an overarching sense of convergence on *the human* that need not otherwise result from the default pattern of convergences taking place in contemporary science and technology. Indeed, conserving humanity's integrity in the face of various induced convergences has become an explicit policy goal, especially amongst EU policymakers, who create distance from US CT initiatives by accusing them of promoting 'transhumanism', which of course the US adamantly, and with *some* justification, denies. Of the two main authors of the original NSF document, only Bainbridge is clearly the transhumanist, with a long-standing interest in Kurzweil-inspired 'cyber-immortality' (Bainbridge 2005), whereas Roco (2002) is more focussed on the reorganising of a 20th century scientific labour force for a 21st century world.

Indicative of the countervailing forces that the CT agenda places on the concept of the human is a set of neologisms introduced by Nikolas Rose (2006), the Neo-Foucauldian sociologist who has coordinated the European Science Foundation's 'Neuroscience and Society' network centred at the London School of Economics:

1. *Biological citizenship* concerns the new ways in which we are coming to relate to each other by virtue of possessing overlapping genomes that are subject to common regimes. Contrary to an earlier ideology of biological determinism associated with the eugenics movement, we are now entering an age in which people will be expected to know, and hence held responsible for, their genetic constitution.
2. *Neurochemical self* refers to the ways in which the parameters of human identity, including our most intimate thoughts and feelings, are coming to be defined in terms of states that are increasingly manipulable by pharmacological or surgical means. This is not quite reductionism because these developments occur at multiple levels of intervention that do not reflect a consistent ontological framework.
3. *Somatic expertise* is a form of knowledge that has emerged to mediate biological citizenship and the neurochemical self by extending regimes of self-management from diet, exercise and regular medical check-ups to periodic cognitive and physical 'upgrades' by means of drugs or surgery. In this context, genetic counselling is an emerging field that envisages our bodies as long-term investment prospects.
4. *Biocapital* captures at once the radical functionalisation and commercialisation of our bodies, which has been greatly facilitated by the biological and technological feasibility of 'xenotransplantation', that is, the successful transfer of organic material – often genetic – from one species to another. The free mobility of biocapital serves to undermine the norm of bodily, and even species, integrity in ways comparable to the role that free trade policies have played in eroding the legitimacy of the nation-state.

I shall return to the transhumanist challenge in section five of this chapter.

4 CT's fixation on nanotechnology: The resurgence of the chemical worldview

CT's fixation on nanotechnology is best seen in terms of the quest for the most finely grained level of reality at which humans can strategically intervene to re-engineer themselves and their environments. A historical frame of reference is provided by the medieval alchemists, who spoke of '*minima materia*', which is sometimes mistranslated as atoms, or ultimate units of matter. In fact, the alchemists were seeking the smallest bits of matter *that retain their functional properties* – largely in the context of medical practice. Homeopathy continues this tradition, especially if one thinks of the serial dilution of toxic materials as a crude prototype of the scaling down of somatic interventions to the nano-level.

However, as one might imagine, precedents from alchemy and homeopathy did not bode well for nanotechnology's early acceptance. The April 1996 issue of *Scientific American* debunked nanotechnology as the latest science hype for promising self-cleaning surfaces, etc., capable of undoing with artifice all the effects that nature had wrought over many years, perhaps even millennia (Stix 1996). In a debate initiated by *Wired* magazine in response to this article in November 1997, Brad Cox, a computer scientist who popularised the idea of 'superdistribution' (i.e. a peer-to-peer tracking system for the spread of digital goods without overarching copyright protection), defined nanotechnology as a 'faith' defined by the premise, 'whatever evolution can do, design can do better' (Cox 1997). He elaborated the point as follows:

The spontaneous orders emerging from evolutionary interaction of autonomous distributed agents with their environment can be improved on by that centrally planned activity the engineering community calls design. Cox argued that the nanotech faith was the death rattle of the 19th century mechanistic worldview, which was inclined to take its models literally, and hence viewed the formation of molecules as akin to the gluing of billiard balls, all in defiance of 20th century knowledge about quantum mechanical effects. At a more general level, argued Cox, the nanotech engineer mistakenly locates himself outside the system he is trying to design, thereby falling foul of evolutionary biology's insights into sustainable environments. Cox himself backed the briefly fashionable 'bionomics' movement, which viewed the economy as an ecosystem that mimics the natural world in a sense aligned to the 'social construction of reality', where 'social' is understood in the distributed micro-sense favoured by phenomenological sociology and Austrian economics (Berger and Luckmann 1968). Bionomics-related research was seen as being conducted by the simulations of 'complex adaptive systems' performed by the Santa Fe Institute.

This early critique cast the enthusiasm for nanotechnology – which at the time was more strongly supported by applied than basic scientists – in terms of the ideology of 'central planning' so favoured by social engineers in the past. Thus, the 1990 book *Bionomics* was largely devoted to evolutionary arguments that undermined Keynes-inspired metaphors for the acting on the economy as 'pump priming', 'cooling down', 'putting on the brakes' and (in the case of corporations) 're-engineering', as if a central planner could do such things without generating long-term, potentially negative unintended effects as well – the economic equivalents of waste and pollution (Rothschild 1990).

However, the prospect of resurrecting the idea of the planned economy, symbolically killed off with the fall of the Berlin Wall in 1989, was not the only target of this assault on nanotechnology. Perhaps more strongly implicated was the proposal put forward from within the free-market capitalist camp by George Gilder, an economist and Republican Party speechwriter who originated 'Reaganomics'. In 1989 he published the best-seller *Microcosm*, pointing to nanotechnology as capitalism's final frontier, now that we are (allegedly) on the verge of acquiring God-like mastery over the fundamental forces of nature. Gilder (1989) thus predicted a nano-cornucopia whereby we could finally realise humanity's biblical entitlement to bring order and prosperity to Earth.

Gilder had in mind this often-cited quote from Eric Drexler's *Engines of Creation* (1986): 'Coal and diamonds, sand and computer chips, cancer and healthy tissue; throughout history, variations in the arrangement of atoms have distinguished the cheap from the cherished, the diseased from the healthy. Arranged in one way, atoms make up soil, air, and water; arranged another, they make up ripe strawberries. Arranged one way, they make up homes and fresh air; arranged another, they make up ash and smoke.' Partly from the proceeds of *Microcosm*, Gilder soon thereafter co-founded Seattle's Discovery Institute, which most notoriously promotes intelligent design as an alternative to Darwinian evolution but has been more practically engaged with the provision of alternative energy solutions for the Pacific Northwest. Gilder himself remains very interested in NBIC-style CT, having played host to Ray Kurzweil at the Discovery Institute where he gathered intelligent design theorists to discuss Kurzweil's proposition that we are 'spiritual machines' (Richards, J. 2002).

Note that nanotechnology's stress on the 'functional' is an anthropocentric concept that presumes an understanding of the arrangement and movement of matter in terms of their instrumentality in bringing about humanly relevant ends. Because the general history of science tends to be told through the history of physics, it is common to treat scientists who persisted in the modern era to regard relations of Newtonian mass and force in purely functional terms – say, as 'energy' – as having been conceptually mistaken. Thus, Joseph Priestley, the polymath chemist who first experimentally isolated oxygen in the 1770s is not normally credited with its discovery because he thought he had invented a technique for purifying air and water (which of course oxygen does), not a fundamental element of nature. Indeed, a convenient way to distinguish the histories of physics and chemistry in the 19th and 20th centuries is that chemistry retained this concern for *minima materia*, whereas physics gave it up in favour of a search for ultimate units as such, regardless of their functional character. Indeed, the rise of the Copenhagen Interpretation of quantum mechanics in the 1920s suggested that ultimate physical reality eludes any ordinary sense of causation. To be sure, nuclear fission, an outcome of physics' search for the ultimate units of matter, proved an innovative basis for both maintaining and destroying civilised life by exploiting properties of matter that can only be called, respectively, 'pre-' and 'anti-' functional. In contrast, CT aims to return science squarely to the functionalist fold.

In the second section, I observed that much has been made of the emergence of nanotechnology as a re-branding exercise for chemistry. This discipline first lost ontological status at the start of the 20th century, after having been reduced to atomic physics, and by the end of the 21st century had lost its sociological status – albeit this time alongside physics – as enrolments dropped and departments closed in the first world. At the dawn of the 20th century, chemistry and physics were on equal epistemological and ethical footing as sources of general natural-philosophical worldviews. At the public level, the differences between physicists and chemists appeared incommensurable: the former concerned with the pure and the latter the practical. However, they also conducted a protracted battle over the reality of atoms, which the chemists denied (except as a theoretical fiction) but the physicists eventually proved, with Einstein's explanation of Brownian motion. After the 1905 discovery, chemistry was increasingly seen as the branch of physics that deals with complex molecules and their applications.

The difference between the physical and chemical worldviews may be summarised in Table 3.1, inspired by Fuller (2000b: 111):

Table 3.1 The Physical *vs.* the Chemical Worldviews

World view	Physical	Chemical
Aim of science	Discover the ultimate nature of things	Construct the most efficient means to our ends
Epistemology of science	Realist	Instrumentalist
Ideology of science	Professional	Industrial
Theory of matter	Atomic	Energetic
Theological horizon	Divine design	Faustian potential

The physical and chemical worldviews can be regarded as complementary, especially from a theological standpoint. The physical worldview draws a strong distinction between divine and human capacities predict and control nature. We aim to discover that beyond which we cannot turn to our own advantage. In contrast, the chemical worldview, much more heretically, imagines humans playing, if not replacing, the divine creator. Here matter is treated not as an insuperable barrier but raw material to be moulded – with more or less difficulty – to serve human needs. What matters is not the ultimacy of matter *per se* but its moment of ultimate plasticity, the so-called edge of uncertainty that the nano-scale promises to provide.

This shift from the physical to the chemical worldview has profound metaphysical implications. Before the 20th century, it was common to distinguish 'natural' and 'nominal' kinds, i.e. things identified in terms of what they are *vs.* what we name them, a Biblical distinction that in its modern form is due to John Locke's adaptation of Thomas Aquinas. 'Nominal kinds' were said to be arbitrary because the things assigned the same names would not necessarily share anything deeper (or 'essential') than our interest in treating them the same. In that sense, all kinds are at least nominal and the question is whether they are natural as well. (Locke shifted the burden of proof to those who claimed to have named natural kinds.) However, by the end of the 20th century, this rather sharp distinction between natural and nominal kinds yielded to more fluid distinctions based on the degree to which we can bend things to our will. Hence, Roy Bhaskar (1975) wrote of the difference between 'transitive' and 'intransitive' dimensions of reality, and Ian Hacking (1998) of 'interactive' *vs.* 'indifferent' kinds, which in both cases roughly corresponded to the objects of the human *vs.* the natural sciences.

Now, however, it may be more appropriate to distinguish between *virtual* and *real* kinds, the latter understood as multiple realisations of the former. The most obvious philosophical precedent here is Gilles Deleuze, who in turn drew on the work of Gilbert Simondon, who held the chair in psychology at the Sorbonne in the 1960s, when Deleuze wrote *Difference and Repetition* (Deleuze 1994). Simondon theorised individuation (i.e. the process of by which one becomes an individual) as products of epigenesis (i.e. the process by which an organism's generic potential is realised in environmentally specific ways), thereby accounting for how, genetically speaking, near-identical members of a given species can come to live such different lives. This marks a radical shift in the ontological focus of scientific inquiry. In particular, 'nature' is cast as only a subset of all possible realisations (i.e. only *part* of the 'real'), as opposed to something inherently 'other' or 'independent' of whatever humans might name or construct. Once again this perspective is familiar from the chemical worldview, in which, say, the difference between 'natural' and 'synthetic' fibres lies entirely in the history of their production and their functional properties, but not in terms of the metaphysical priority of one to the other, since both the 'natural' and the 'synthetic' are composed of the same fundamental stuff – and the latter may indeed count as an *improvement* over former. By extension, 'mind' and 'life' lose the metaphysical mystique associated with their natural origins and come to be assessed simply in terms of the properties possessed by their realisations – be they human, carbon-based, silicon-based or some cyborgian mixture. I shall pick up this point in section six's discussion of 'ableism'.

Starkly put, in this third metaphysical phase, a thing's identity is no longer constrained by its history, not even its Darwinian evolutionary history. Thus, as we get better at pharmaceutically manipulating genetic expression and neural circuitry with an eye to long-term improvements – be it through direct incorporation into the next generation's genetic potential or less directly through regular corrective medical interventions (cf. vaccinations) – the more hollow the concern voiced in the following article will appear:

Human enhancement beyond evolution

'If it is such a good idea, why has evolution not built us that way?' That is the question two philosophers say we must ask before we attempt to enhance our human capabilities.

> We already augment our minds with drugs such as Ritalin and Modafinil, our sexual performance with Viagra and our immune systems with vaccines. These are nothing compared with what might be on the way, from brain implants for a better memory to genetic modifications for sports performance (*New Scientist*, 13 May 2006: 32).
>
> Before we consider forging ahead with these technologies, we need to consider why we haven't already evolved that way, say Nick Bostrom and Anders Sandberg of the Future of Humanity Institute at the University of Oxford. This will allow us to identify when it is feasible for us to outdo nature, they say, and when it is not.
>
> Before anyone considers giving humans greater brain power, for example, they should first show that the only reason we don't already have more mental capacity is that the resulting energy demands would have been a disadvantage for our hunter-gatherer ancestors when food was scarce. Now food is more plentiful, it might be OK to forge ahead, but if there is no convincing guarantee that this enhancement no longer poses a problem, it might be wiser to steer clear of it. 'The human organism is enormously complex,' says Bostrom. 'If we go in blindly and change things at random, we are likely to mess up.' He presented the idea last week at the Transvision conference in Helsinki, Finland.

I highlight this short article, which appeared in the 26 August 2006 issue of *New Scientist* because caution with respect to human enhancement policies is being urged on evolutionary grounds from a most unlikely source, namely, two intellectual leaders of the transhumanist movement. It would seem that even transhumanists – at least the academically respectable ones – continue to trade on an old rhetoric of evolutionary 'anchoring' that harks back to a time – from the late 19th to the late 20th centuries – when the ancient ancestry of our genetic traits (e.g. vestiges of the 'reptilian' or 'primate' brain) was associated with their relatively strategic impermeability. In the philosophy of biology, this perspective is associated with the 'Weismann Doctrine', named for the German embryologist normally credited with experimentally demonstrating the impermeability of an organism's second-order 'germ' cells (aka its genetic potential) to environmentally induced change in its first-order 'somatic' cells (aka its actual embodiment). In short, what happens to the parent normally does not happen to the child, unless the parent had already passed on the capacity to register the traumatic experience. Of course, by the early 20th century, it was generally granted that irradiation, strictly speaking, violated the Weismann Doctrine by greatly multiplying the capacity for genetic mutation. Unfortunately, it did not do so in a strategically tractable way, as, say, followers of Lamarck would have liked. Many science fiction B-movies produced in the early years of the post-World War II 'Atomic Age' reflect this disjunction. However, CT precisely revisits the Lamarckian dream with better science.

But as a matter of fact, as transhumanists would be the first to point out, we are gradually discovering ways of re-engineering processes and properties that originally developed over millions of years. Even from an evolutionary standpoint, there is no reason why biological traits that have been around for aeons cannot be successfully changed overnight, provided the presence of environments where individuals possessing the new traits prove 'adaptive' (i.e. reproduce themselves). To be sure, this is much easier said than done. Indeed, the extreme prospects of genetic and neural re-engineering – both in terms of risks and benefits – revisit the classic questions of social engineering. However, addressing them adequately has less to do with respecting the deep past than with reconstructing today's socio-technical world to render it hospitable for any such biologically modified beings. The nostalgic appeal to an evolutionary naturalism simply obscures what is, in effect, a straightforward political decision about the care with which we project future generations. Letters to the editor about the *New Scientist*

Table 3.2 The Metaphysical History of Genetics

	Metaphysical distinction	**Genetic orientation**	**Capacity for intervention in life processes**
Before 20th century	Natural *vs.* Nominal kinds	Linnaean species creation	Minimal: Fundamental life processes out of human hands
20th century	Intransitive *vs.* Transitive kinds	Mendelian population genetics	Selective breeding can affect later generations
After 20th century	Virtual *vs.* Real kinds	CT-style nano-bioengineering	Alternative realisations of genetic potential possible in same generation

piece reflected critically on the transhumanists' continued normative reliance on evolution. One observed, quite properly, 'Evolution didn't "build" us at all. It can only play the hand mutation deals it. If no mutation occurs giving rise to a particular characteristic, no matter how much of a "good idea" that characteristic is, it will not arise. We, however, have the capacity for foresight and so can fine-tune some of evolution's less elegant solutions.'

Table 3.2 encapsulates the foregoing three-stage metaphysical transformation in what kinds of things there are is to correspond them to the three main phases in the history of genetics, with CT bringing the final stage to fruition.

5 Better living through biology: A hidden theme in the history of social science

Despite considerable controversy surrounding the term 'human enhancement' as a goal of CT, with the EU in equal measures suspicious and sceptical of US aspirations, nevertheless such disagreements are less over the desirability of enhancement *per se* than the form it takes. As we have seen, 'enhancement' promises that individuals will enjoy greater consumer choice but also longer economic productivity, thereby enabling lessening state welfare burdens. It would seem, then, that there is something for everyone across the political-economic spectrum.

There is a long history of treating genetic variability in competitive terms, as played out over successive generations of socially delineated 'races', 'clans' and 'families'. The interest in enhancing human performance is ultimately rooted in the palpable differences in achievement that emerge from examining these various lines of human descent. In particular, those from modest origins often pick themselves up but never reach the top without violence, and then only temporarily, whereas those who start on top often regress to a position of mediocrity if not outright degeneracy, unless they prove to be of sufficiently strong 'character'. However, it has been long thought that some targeted intervention might be able to alter both these tendencies – notably the first major work of Western political philosophy, Plato's *Republic*.

While most subsequent theories of politics have concentrated on preventing the rot from setting in (e.g. through constitutional checks and balances and various incentives to prevent corruption), Plato was distinctive in trying to raise the bottom by identifying promising offspring from all classes and subjecting them to special training over the course of several decades to enhance their latent potential for leadership. If Freud held that a child's future was sealed by age five, Plato held that it was around that age that the child's nascent responses to the world could be channelled for maximum social benefit.

Though lacking anything like a modern theory of genetics but possessing a keen sense of Greek history, Plato was struck by the unreliability of family background as a predictor of desirable qualities like leadership. Nevertheless, he believed that a stable social order requires just such a belief in the heritability of achievement. The value of heritability lay in the security one feels from anticipating what people are likely to do under normal circumstances, given their past, which then allows for their acts to be encouraged or prevented. Plato spoke of this as a 'noble lie', the so-called 'myth of the metals', the quasi-racist, caste-like basis of a stable social order, which justified segregating the best from the rest. However, this folk theory needed to be supplemented by a more esoteric theory that recognised the inevitable uncertainties that resulted from people of perhaps a fixed genetic make-up encountering circumstances, themselves perhaps separately predictable, but beyond the control of those encountering them.

The big difference between how Plato and we think about the prospects for human enhancement is that unlike Plato, who conceptualised the issue in terms of decisions taken about individual lives, the CT agenda operates at two steps removed, selecting research trajectories likely to result in enhancement innovations that, at least in principle, would be available to the full range of inhabitants of the nations promoting the CT agenda. To be sure, which particular individuals end up benefiting from these innovations is left open in a way Plato would not approve. To a large extent, this difference in approach reflects Plato's greater certainty about the consequences of his decisions. He believed that the requisite knowledge was already available but that people were normally too self-interested to be trusted to make the right decisions. Thus, Plato established the Academy as a school for aspirant philosopher-kings, who would be trained to adopt the universal standpoint as their own default basis for taking decisions. To be sure, Plato regarded this as a difficult task, requiring several decades of matriculation – but *not* the commission of specialised research.

Plato's folk theory of the heritability of achievement, the 'myth of the metals', was revisited with new empirical vigour in the late 19th century by Darwin's cousin, Francis Galton, who coined the term 'eugenics' for the project of tracing family lineages in order to identify, and cultivate, lines of achievement (Kevles 1985; Renwick 2012: chap. 2). This project was politically attractive to an emerging liberal-socialist sensibility, associated with the Fabian Society in the UK, that on the one hand was keen to remove the hereditary privilege of the House of Lords, which typically rested on the achievement of one ancient ancestor who turned out to have been an exception in a family history whose members have regressed over successive generations; and on the other hand, feared that the advent of majoritarian democracy would swamp the efforts and aspirations of the talented unless they reproduced themselves in sufficient numbers.

Although the underlying theory of genetics changed radically over the 80 or so years that saw the likes of Galton, Karl Pearson, Ronald Fisher and Julian Huxley advance versions of what is often called 'positive eugenics' (as opposed to the 'negative eugenics' associated primarily with culling, as practiced *in extremis* by the Nazis), they all agreed that not everything was worth preserving in the human gene pool simply because the gene pool was 'human'. In this respect, these thinkers accepted the premise of all versions of modern evolutionary theory, namely, that species are not fixed essences (e.g. specially created by God) but mutable sites for the collection and transmission of genetic material.

The history of eugenics is relevant to the project of human enhancement because it establishes the point-of-view from which one is to regard human beings: namely, not as ends in themselves but as means for the production of benefits, be it to the economy or to 'society' more diffusely understood. The Abrahamic or Kantian idea of humanity as a species-being in possession of its own unique integrity and autonomy (aka 'dignity') is largely relegated to ethical 'side constraints' for the conduct of research and 'precautions' related to anticipated negative consequences of such research and its applications. Such a 'posthumanist' position most strongly resembles what happened to the idea of *producer* in classical political economy. In authors from Aquinas to Locke, a 'producer' was the worker through whose creative transformation value was given to nature. It was associated with humanity's spark of divinity. However, by the early 19th century, 'producer' had come to name the workplace manager whose organisation of workers enabled the efficient flow of goods and services. In other words, a producer became a human whose job was to transform other humans, as if they too were simply part of nature. An awareness of this semantic transformation lay behind Marx's early critique of capitalism, especially in terms of the alienation of the worker from his labour as the abstract factor of 'productivity' that requires the supplementation, if not outright replacement, of people with machines and other artificial arrangements. This self-alienation of the mental and physical parts of production was crystallised in the 20th century through various theorisations of an intellectually driven 'managerial class' that would run a firm like an army – from 'the top', which was exemplified before and after the Second World War by, respectively, the ex-Marxists James Burnham and Karl Mannheim. The model had been already provided in the specialist training of the French *grandes écoles*, which Hayek (1952) held to be responsible for all perverse modern applications of science as a technology of radical social transformation.

The CT agenda, especially in the NBIC form promoted by 2002 NSF document, harks back to this early understanding of social science, one that predates the field's separation into distinct disciplines or, for that matter, its clear differentiation from the natural sciences. It is a vision most recognisable as Auguste Comte's original version of 'sociology' as the overall development of science brought to self-consciousness, as humans are finally incorporated as proper objects of scientific inquiry, thereby providing the site for the integration and collective self-governance of the all the sciences. Convergence on the ideal social order on a global scale would presumably soon follow. A slightly less grandiose, less theoretically freighted and more policy-oriented precedent of this vision actually came close to the horizons of today's CT agendas. I have in mind the 1814 proposal of Comte's mentor, Count Henri de Saint-Simon, *The Reorganization of European Society*. Saint-Simon held that regardless of Napoleon's personal fate, he had succeeded in consolidating Europe as an idea that could be taken forward (by others) as one grand corporate entity, to be managed by a scientifically trained cadre, modelled on the civil engineers trained in the *Ecole Polytechnique* (Hayek 1952: chaps. 12–16).

A striking feature of Saint-Simon's vision, relevant for our purposes, is his generalisation of Adam Smith's hostility to the barriers that owners, and laws governing ownership, placed to the productive use of capital. The form of capital Smith mainly had in mind was land, whose owners could derive income by charging rents for simple occupancy. Saint-Simon's CT-relevant innovation was to propose that *ownership of one's body* was the main barrier to increased productivity – what is now euphemistically called 'underutilised human capital'. By analogy, Saint-Simon objected to the idea that individuals, simply by virtue of self-possession were entitled to certain basic goods. To be sure, by the late 18th century, ideas of liberty as an 'inalienable' right premised on the 'dignity' of the person had become the standard by which political regimes were judged, on the basis of which the American and French Revolutions were justified. And in this respect, Saint-Simon was a 'counter-revolutionary' thinker. However, from the standpoint of CT, he was ahead of his time.

The radical assumption behind Saint-Simon's proposal was that possession does not entail competence. Property ownership had been traditionally required for political participation because it was presumed that owners must be able to manage their holdings effectively in order to thrive: i.e. they displayed on a small scale the sort of judgment required on a large scale. This line of reasoning was extended to self-ownership in the late 18th century to incorporate tradesmen and professionals who may not be landholders but whose gainful self-employment revealed their competence. Saint-Simon's proposal gave a perverse spin to this development by shifting personal competence from an 'input' to an 'output' measure – i.e. from presumptive possession to revealed productivity. In short, Saint-Simon legitimised the idea that, on a show of competence, not only might political power be granted to those who previously lacked it (such as tradesmen and professionals) but also the converse applied, such that delinquent landholders might lose the right to dispose freely of their property. He notoriously made the point by arguing that France would lose its civilisation and prosperity if it lost its scientists and artists, but nothing would change if it lost its priests and aristocrats. It was this assessment that led Marx to deride the *rentier* class for its promotion of 'rural idiocy'.

The 19th century made the shift to Saint-Simon's perspective increasingly plausible as the state came to represent society as a corporate 'national' entity with a life and purpose above and beyond those of its constitutive individuals. The administration of this corporate entity was entrusted to a bureaucracy – whom Saint-Simon envisaged as consisting of industrialists and technocrats – with the power to redistribute the nation's wealth so as to ensure maximum productivity. Recall, once again, that 1814 was before the natural and social sciences were clearly distinguished. This bears on what 'redistribution' might have meant. It is now easy to imagine Saint-Simon as having been concerned with redistribution *only* at the level of material wealth, i.e. with the state's ability to tax and spend. However, he was also interested in the redistribution of 'sentiment', largely through changes in what, after Claude Bernard, came to be called the 'internal' (i.e. the organism's physiology) and 'external' environments responsible for their generation and maintenance. As we shall see below, this aspiration establishes his relevance to the 2002 NSF report.

Saint-Simon – and certainly Comte and sociology's academic founder, Emile Durkheim – saw the matter in terms of 'moral education', which in practice meant a reprogramming of each generation's brains to undo the misconceptions (or 'ideology') instilled by religious instruction, not least the idea of a mental life independent of both the natural and social order, the so-called seat of the soul, the pseudoscience of which was 'psychology'. While these thinkers thought of reorienting brains to society largely in terms of altering the 'external environment;' they certainly aspired to intervene more directly in the brain. Indeed, an often neglected feature of 19th century debates over the foundations of the social sciences – then often called the 'moral sciences' – is the enthusiasm for a positivistically upgraded science of *medicine* to become the basis for a unified policy science that might pass for 'sociology'. CT, especially in its NSF guise, should be seen as revisiting this prospect at a time when the differences between the natural and social sciences – not least the biology/sociology interface – have begun to lose their institutional and intellectual salience.

Here it is worth observing that the biology/sociology interface remained porous as long as the so-called the Weismann Doctrine was not in effect. In other words, as long as biologists found no reason to think that physical changes to a current generation of organisms would have long-term effect on offspring, it became convenient to distinguish biology from sociology in terms of a focus on genotypic *vs.* phenotypic changes – the former change bearing on the latter, but not vice versa. To be sure, the Weismann Doctrine is alive and well amongst evolutionary psychologists who explain the limited variance of human socio-cultural responses to their physical environment in terms of genotypic anchoring. However, the promise of CT's capacity to switch genes on and off and otherwise produce permanent effects on the genome in a single generation suggests the resurgence of a sensibility closer to Saint-Simon and Comte, both of whom were sympathetic to Lamarckian views of evolution. Following recent analytic philosophy of mind, including Fuller (1993) and Turner (2007), we may distinguish four modes in which sociology might relate with neurophysiology:

(1) *Dualism* – they describe two relatively autonomous domains, perhaps because of the Weismann Doctrine (this has been sociology's default position for most of the 20th century, but CT-driven prospects of neuro- and even geno-plasticity increasingly make this option untenable);
(2) *Eliminativism* – the position of the French positivists, whereby 'psychology' is just a false religiously inspired theory of how brain-society interactions work;

(3) *Reductionism* – different states of social being (e.g. a secular ideology and a religious belief) are reducible to common brain patterns;
(4) *Functionalism* – different brain patterns converge on a common state of social being (e.g. multiple constituencies for a political party or multiple markets for a product).

In its pre-scientific 'therapeutic' mode, medicine was largely concerned with preparing 'patients' – literally passive beings – as they pass through the natural course of their lives. However, the 19th century came to see infirmity and death as enemies of the body politic to be overcome through regular and systematic medical treatment, functioning as a kind of micro-level national security system. This change in sensibility is normally attributed to the late 18th century physiologist Xavier Bichat, who figured as a major saint in Comte's positivist revision of the Roman Catholic Church's holy calendar, despite the fact that he was dead by age 30. As mediated by the founder of French experimental medicine, Claude Bernard, Bichat's idea passed into the work of Durkheim, who quite explicitly treated deviance as moral pathology (Hirst 1975).

Moreover, this view was by means restricted to France. In Germany, Rudolf Virchow as early as 1855 argued for medicine as the scientific basis of the law, calling for medical doctors to function in a proactive capacity, akin to the newly established legal institution of the police. According to this line of thought, warding off disease (especially epidemics) is like warding off crime: Both rob society of its productivity but they differ over the physical level at which the infractions occur, with medical doctors operating at a finer-grained level than the police (Saracci 2001). Carried to its logical extreme, this epidemiological perspective on politics encourages nations to think about themselves as 'always already' in a 'state of permanent emergency', as discussed in the first chapter and popularised in the Steven Spielberg film *Minority Report* (2002). While not sufficient to enable the convergence of the disciplines of medicine and law, this line of thought has continued as, say, the basis for child vaccination campaigns, in which negligent parents can become subject to prosecution. And now we might not be far from the day when the right to give birth requires prior consultation with a genetic counselor who apprises the pregnant woman of both her options and her liabilities for their consequences (Rose 2006: chap. 4).

In short, were he teleported across the two centuries that divide him from us, Saint-Simon could recognise the following slogan, taken from the NSF document, as a more advanced version of what he had advocated. Table 3.3 provides a chart of the relevant translations:

> If the Cognitive Scientists can think it,
> the Nano people can build it,
> the Bio people can implement it,
> and the IT people can monitor and control it
> (Roco and Bainbridge 2002a: 13).

Table 3.3 'Converging Technologies' Before and After the 20th Century

Saint-Simon (early 19th century)	Roco & Bainbridge (early 21st century)	Shift of focus
Social science	Cognitive science	From institution to individual
Carceral institutions and urban/regional planning	Nanotechnology	From external to internal environment
Medicine (both and forensic and corrective)	Biotechnology	More intensive interventions
Vital statistics (administrative sciences)	Information Technology	More extensive data gathering

The applied epistemologist Jean-Pierre Dupuy (2004) has argued that a unique feature of the nano-driven character of the CT agenda is that proposals have been made for the normative regulation of scientific research – sometimes resulting in explicit guidelines – long before such research actually exists, let alone has borne socially relevant fruit. But Dupuy's claim is strictly speaking false. An important earlier precedent is the 'anticipatory governance' of alchemy by the Roman Catholic Church, especially after the Papal Bull of 1317, which prohibited the project of transmuting base metals into gold on both moral and economic grounds: morally, it arrogated to humans what properly belonged to nature, and economically, it threatened to upset the exchange value of precious metals. Analogous concerns about the destabilisation of nature and the economy are raised today about nanotechnology, especially in light of the claims of its more zealous advocates like Drexler and Kurzweil. However, as was the case 700 years ago, the practices that would require such governance have yet to materialise. The updated version of the alchemist's 'philosopher's stone' would be the creation of an assembler at the nano-level that can then assemble other nano-molecules.

The point to underscore is that the Papal Bull was announced without any evidence that the more ambitious elements of the alchemical project were even close to realisation, this despite the hype generated by the Oxford Franciscan Roger Bacon, the medieval answer to Drexler, who believed that alchemy is part of humanity's Biblical entitlement as having been created in *imago dei* (Noble 1997). In any case, this 'anticipatory governance' orientation has become the main framing concept of the largest social science initiative associated with the US CT agenda, the 'Nanotechnology in Society' network centred in Arizona State University, under the leadership of David Guston and Daniel Sarewitz. It would seem natural to translate a concept like anticipatory governance into the language of ethics, perhaps as an extension of the 'precautionary principle' used in ecological discourses. However, this fails to capture the proactive character of the lines of inquiry pursued under the concept, which more strongly resembles public relations or even marketing.

Two aspects of these 'anticipatory' activities are relevant here, one from the science side and the other from the public side. First, practitioners of certain branches of materials science and chemical engineering – if not chemistry more generally – have increasingly identified their field of research as 'nanotechnology'. This has enhanced the sense of forward momentum to nano-driven fields in citation indexes that depend on self-characterisation for their keywords (Porter et al. 2008). Second, social scientists in both the US and EU have been interested in not only surveying public opinion on current developments in nanotechnology but also anticipating the reception of future nano-based products. Cynthia Selin has been very much in the forefront of these developments, starting with her PhD in knowledge management from the Copenhagen Business School, which grounds this entire approach in the classic Cold War foresight methodologies that used computer simulations to anticipate in aid of preventing conditions of attack. In the post-Cold War period, the intelligence gathered from defectors has been arguably replaced by a more pro-active 'participatory design' strategy that invites those potentially impacted by nanotechnology to voice their concerns, which may be then addressed as a design feature of the technology (Selin 2007). Intentionally or not, this serves to acclimatise citizens, in the company of their peers, to whatever nano-driven changes might be on the horizon, thereby updating the concept of 'self-fulfiling prophecy'.

These 'nano-futures', which are presented both live in 'science cafés' (i.e. the American version of the 'café scientifique') and in cyberspace through wiki-media. The scenarios are initially vetted by the relevant scientists so as to be sufficiently plausible for people to take seriously. In social psychology, this strategy is often dubbed 'inoculation', the suggestion being that by allowing people to spend time thinking and talking about extreme or pure cases of some potential threat, you have laid the groundwork for the acceptance for a less virulent version. At the very least, you have normalised the idea in their minds. Of course, at the same time such scenarios lower one's guard to the potential harms caused by nanotechnology, they also raise one's expectations that its social benefits are forthcoming. But this too may be interpreted along Janus-faced lines: The anticipatory acceptance of nanotechnology may lead, on the one hand, to an anti-science backlash if sufficient benefits are not forthcoming or, on the other, to a willingness to interpret all manner of marginal nano-driven improvements as indicative of greater things to come.

For Dupuy, these nano-futures are high-tech versions of the performative, or 'self-fulfiling', character of prophecy, whereby a notional preference for a certain future, which the prophet channels as the voice of God or the scenario elicits from the participants, serves as a groundwork for what in retrospect will enable people to say that they were prepared for what eventually happened. Of course, strictly speaking, self-fulfiling prophecies need not turn out to be true but the import of taking the prophecy seriously is to think in terms of tendencies in the present that would indeed be responsible for the prophecy coming true, were it to come true. Similarly, as people become accustomed to thinking in terms of nano-futures, while the relevant scientific breakthroughs that would turn these scenarios into realities may not happen any more quickly, people will be primed – and inclined to provide further groundwork (in terms of funding, 'anticipatory governance' regimes, etc.) – to recognise and incorporate the realisation of the nano-futures when (and if) they happen.

One feature of this 'priming' of the future is worth highlighting, as it bears on the transhumanist futures that, as we shall see in the next section, some enthusiastic bioethicists have begun to project. The historic appeal of Lamarck's theory of evolution lay in the prospect of improving oneself through deliberate effort, the results of which would have continuing genetic consequences. The panoply of proposed CT-based enhancement strategies promise to deliver on at least this part of Lamarck's vision. However, the justifiability of this optimism depends on how one identifies the nature of the relevant interventions.

Bioethicists and others hoping for a Neo-Lamarckian revival tend to talk about genes as a population geneticist would, namely, as bearers of socially significant traits – as if that captured the character of our interventions in the genome. Thus, thought experiments to test our intuitions about the morality of enhancement typically go like this: 'Suppose a treatment was available to switch on a gene that would enable your child to cognitively mature at such a rate that he could avoid primary school altogether....' The problem with this scenario is *not* that no one currently faces such a problem but rather that progress in our ability to intervene at the nano-level of life – and to monitor the relevant consequences – is best understood in terms of how molecular biology thinks about the gene, which has to do with the propensities of various protein configurations in a given biotic environment, such as the human body. As the leading historian of the field put the matter:

> How is gene defined: population geneticists follow traits, whereas molecular biologists follow protein: 'for the molecular biologist, a gene is a fragment of DNA that codes for a protein. For a population geneticist, it is a factor transmitted from generation to generation, which by its variations can confer selective advantage (positive or negative) on the individuals carrying it' (Morange 1998: 249).

A similar dichotomy arose between the relatively cautious Max Delbrück and his intellectually more adventurous mentor Erwin Schrödinger, who together facilitated the passage of a generation of physicists and chemists into the field that is now known as 'molecular biology'. The distinction between their two views is summarised in Table 3.4, which is inspired by

Table 3.4 Genetics Before and After the DNA Revolution

Genetics *vis-à-vis* the DNA Revolution	Before	After
Species of reductionism	Mendelian distribution of traits	Monodian architecture of proteins
Physics patron	Delbrück	Schrödinger
Physical model of the gene	Force-like	Mass-like
Precedent from history of biology	Preformation	Epigenesis
Moss's distinction	P-gene	D-gene

Rosen (1999: chap. 1). The biochemist-turned-philosopher Lenny Moss (2003) has epitomised their differences in terms of alternative conceptions of the gene: P-gene (a gene for a specific trait) and the D-gene (a gene as a potential that can be actualised in many different ways).

In short, we may be getting better at, say, gene switching or brain boosting but our social categories do not naturally map on either the causes or the consequences of such interventions. We are basically just learning how to manipulate our proteins better. In this respect, a society that encourages the study and application of CT-oriented research is forced to conceive of the activity as an opportunity to use our own bodies as sites for biomedical experimentation and bioprospecting. I say this not to discredit the transhumanist ambitions but to highlight the attendant changes in the sense of self – as well as our relationship to others – in what amounts to a scientific license for risk-seeking behaviour of the most fundamental order. My guess is that transhumanists routinely commit this category mistake because they are so keen to demonstrate the feasibility of overcoming traditional 'natural' boundaries by artificial means – even, so it seems, these means are sociologically speaking either irrelevant or deleterious.

6 The transhumanist challenge: Can CT 'enhance evolution'?

John Harris, editor of the *Journal of Medical Ethics* and Professor of Bioethics at the University of Manchester School of Law, is probably the most intellectually challenging moral philosopher writing in Britain today. He has recently published *Enhancing Evolution* (Harris 2007), based on a series of lectures given at Oxford's James Martin Institute for Science and Civilisation in 2006, presents the most systematic case to date for the value of artificially enhancing the human condition along broadly CT lines. Although Harris does not explicitly endorse a 'transhumanist' ideology, he admits that the liberal policies he supports on enhancement may eventually result in a species-change that might be properly called 'transhumanist' (Harris 2007: 37–8). One is reminded here of the back-door route to socialism from capitalism through an enlightened sense of self-interest that recognises the long-term benefit of a progressive income tax regime to productivity and hence prosperity. Like socialism, transhumanism retains an air of political incorrectness that requires its ends to be achieved by (at least verbally) indirect means.

Some other caveats need to be issued about Harris' argument at the outset. Harris defends 'enhancing evolution' on Neo-Darwinian and utilitarian grounds. However, one might start from Neo-Darwinian and utilitarian premises and project a rather different future from Harris'. In this respect, a conspicuous omission from his otherwise wide-ranging treatment of actual and potential opponents is Peter Singer, the only philosopher whose global influence exceeds Harris' on bioethical matters. Singer shares Harris' starting point but reaches significantly different conclusions. Much more than Singer, Harris takes a liberal-individualist stance towards utilitarianism, as if Bentham were simply a natural extension of Locke. He interprets the utilitarian maxim 'the greatest good for the greatest number' as something for everyone to decide for themselves as long as it does not prevent others from doing the same.

An alternative reading of the utilitarian maxim, one closer to Singer and more in the original spirit of Bentham's maxim, would deal with matters in a more aggregate fashion, in which case one might query the benefit-to-cost ratio of regularly enhancing a deficient individual *vis-à-vis* simply transplanting that individual's remaining functional parts to others who might make better use of them. After all, utilitarianism is, strictly speaking, a philosophy dedicated to the maximisation of *social* welfare, and hence not *a priori* committed to the bodily integrity – let alone indefinite enhancement – of *individuals*, whose value is mainly as sites for registering society's pleasures and pains. Thus, according to Rawls (1971), utilitarianism founders on personal integrity but is this really any different from species integrity: i.e. we're all samples of the same gene pool. Our humanity is that we set boundaries, categories, whereas nature by itself would be entirely indeterminate.

This subtle but important point was brilliantly satirised more than a decade ago by the political theorist Steven Lukes in the novel, *The Curious Enlightenment of Professor Caritat* (Lukes 1996). Lukes envisaged a utopia called 'Utilitaria' a land whose motto was 'From Welfare to Farewell', as citizens came to think of their legacy in terms of the body parts they could bequeath to fitter specimens, once their own bodies exhibited diminishing productivity returns on biomedical investments. It is easy to ridicule such a sensibility, but it actually captures a world in which people have come to realise that they are all made of the same stuff, given some largely accidental marginal differences.

If anything, from a Neo-Darwinian standpoint, Lukes' Utilitaria is much too tame. One could further argue that its regime needs to be extended to all animals, whose genomes after all differ from human ones by no more than 5%. At that point, we enter into Peter Singer's bioethical paradise, which would turn the welfare state into a guarantor of the efficient transfer of genetic material to enable the maximal productivity of the widest range of species (Singer 1993). This would amount to treating genes as pure capital (or 'biocapital', to use Nikolas Rose's term) in search of greater mobility, with humans as just one of its many transient species bearers. (Imagine Richard Dawkins 'selfish gene' vision of evolution implemented as an extension of free trade policy.) In this respect, the molecular revolution has enabled biology to advance more swiftly along the trajectory charted in the 19th century in political economy, during which 'value' came no longer to be seen as ultimately grounded in land or even labour but inclusive of anything that could be exchanged at a price. Similarly, nowadays 'life' is not restricted to naturally evolved life-forms but extended to artificial entities that can function in a life-like fashion, i.e. bearers of biocapital. Given the closeness of natural history and political economy in the 18th century, with figures such as Linnaeus and Buffon having contributed to both fields (the idea of 'ecology' as nature's economy is a remnant of that era), it is striking just how *long* it has taken for life to become fully absorbed into the processes of commodification. Generally speaking, until the mid-20th century's consolidation of the Neo-Darwinian synthesis, biological thought held on to a strong distinction between 'natural' and 'artificial' that political economy had abandoned at least a century earlier.

The nightmare scenario, then, would not be the Marxist one that humans might be replaced by technology once their productivity flags, which would simply leave them unemployed, alienated but at least still alive. Rather I have in mind the Darwinist scenario that particular humans might need to be culled to ensure an efficient division of labour amongst various species (aka symbiosis) in a sustainable ecology. Nazi Germany was the first society that claimed to act on the basis of this principle, which eventuated in the 'culling' of millions of Jews. I have discussed this precedent, including the likelihood that a similar 'culling' might occur in the future diffusely through the aggregation of individual choices in a 'bioliberal' regimes in Fuller (2006b: chap. 14). Moreover, as in the case of Utilitaria, another distinguishing feature of any such diffusely executed culling in the future is that organs and other biomatter would be farmed and harvested from the victims, as already happens (at least from a natural law standpoint) during some forms of stem cell research.

Harris disappointingly fails to come to grips with this alternative future that could easily follow from his own Neo-Darwinian and utilitarian premises. He avoids discussing not only Singer but also more generally animal rights, android rights or, for that matter, any broader ecological orientation – including the physical side-effects of nano-based biotechnologies that in the future may be used, say, to regenerate our organs or cleanse our bloodstream. Harris' ethical universe is resolutely anthropocentric and relatively innocent of concerns about cyborgs or any other witting or unwitting hybridisation of the human condition. However, the most touching feature of Harris' naïvete is his reliance on Darwin's authority.

What makes Harris' faith in Darwin touching is that he retains so much of the unfounded humanist sensibility of Darwin's early followers. Like them, Harris cautiously welcomes transhumanism as humanism brought to self-realisation – not as a fundamental discarding of the human as an altogether inferior form of life. For a glimpse into the limits to Harris' imagination, consider this bland statement:

> It is difficult, for me at least, to see any powerful principled reasons to remain human if we can create creatures, or evolve into creatures, fundamentally 'better' than ourselves. It is salutary to remember that we humans are the products of an evolutionary process that has fundamentally changed 'our' nature (Harris 2007: 40).

Of course, it is difficult – especially if you cannot imagine that those future creatures might *lack* features that are now core to human identity. Here I don't mean creatures lacking in such historically deep human capacities as cognitive abstraction or moral reflection. I mean something much more basic. If the worst scenarios of global warming advocates turn out to be true, then our evolutionary successors might be best adapted to live in a restricted sensorimotor environment, so as to ensure minimal disturbance to the ecosystem. In that case, those whom we now call the 'disabled' may well constitute mutational vanguard of this posthuman species. Their advanced intellects would not be enhanced by capacities to intervene far beyond their physical location. (Think Steven Hawking.) From a strictly Darwinian standpoint, such a prospect must be taken seriously: After all, consider the downsized version of reptilian life that has descended from its dinosaur heyday.

In contrast, Harris, like many of today's so-called secular humanists, still harbours late 19th century hopes that evolution ultimately converges upon humanity's utopian fantasies. Yet, any substantial realisation of those fantasies requires *deviating* from the default trajectory of evolution, at least as conceptualised in Darwinian terms, namely, a process lacking both knowledge and hope of the sort of fine-grained understanding of heredity that now provides *prima facie* plausibility to Harris' arguments for enhancement. In this respect, Lamarck is a surer guide than Darwin – especially in terms of the debates that normatively matter. The difference between Lamarck and Darwin is usually conceptualised in terms of how one explains adaptive variation in nature, with Lamarck allowing for a much greater amount of genetically transmitted learning than Darwin. However, the truly significant difference lies in their alternative conceptualisation of the evolutionary process. Whereas Darwin envisaged the origins of all species in terms of lines of common descent, Lamarck postulated that life was being created from scratch all the time, yet all creatures evolved towards some superior version of humanity.

Thus, Lamarck was much less beholden than Darwin to species' physical morphology as a guide to what they might ultimately become. We tend to forget that, unlike Gregor Mendel, Darwin's belatedly recognised contemporary and the founder of modern genetics, Darwin himself stressed the *disanalogy* between the workings of natural selection and 'artificial selection', that is, the collected practises of animal and plant breeding that have informed agricultural progress over the centuries. Because Darwin believed that natural selection would always trump our best efforts at artificial selection, he was relatively pessimistic about humanity's capacity to relieve the more miserable aspects of our collective existence, other than by inhibiting the reproduction of those suffering from demonstrable genetic deficiency. Harris thus fails to realise that Darwin's true descendants are to be found amongst defenders of the precautionary principle, even as he humorously dismisses then for their extreme risk-averse policy perspective (Harris 2007: 34–5).

Harris' naïve confidence in Darwin's support is exemplified in the 'retro-futurist' image that graces the cover of *Enhancing Evolution*, namely, the flexed arm muscle of a comic book Superman. In the mid-20th century, the phrase 'making better people' did indeed conjure up the idea of beings that were excellent versions of our current selves, as in the case of Superman, whose irradiated body expedited genetic change in generally desirable directions. Much of this popular imagery was based on the work of Hermann J. Muller, a pro-Soviet US geneticist who won the 1946 Nobel Prize in Physiology or Medicine for discovering X-ray mutagenesis in fruit flies (Paul 1998: chap. 2). However, Muller's own considered view was that irradiation usually produced lethal mutations that expedited death not evolution. Nevertheless, this line of thought must be considered as part of the tradition interested in simulating Lamarckian effects by Darwinian means. (One of Muller's fellow-travellers was Conrad

Waddington, who housed Muller at the University of Edinburgh's Institute of Genetics in the early days of World War II, once Stalin's repressive policies made even Muller's eugenics-friendly research unfeasible.) While Muller avoided the transhumanist obsession with expediting evolution, he pioneered the movement's obsession with preserving (nowadays cryogenically) superior genetic stock by stressing how environmental pollution (not least from ambient radiation) was bound to deteriorate the human gene pool. Muller's career, which deserves close study today, highlights the Sisyphean dimension of transhumanism – i.e. unless continually proactive measures are taken, humanity's positive features will be undermined in the long term.

But nowadays transhumanism's normative horizons veer towards what has been called *ableism* (i.e. able-ism), which aims for the indefinite promotion of various abilities, regardless of the species identity of their possessors. A cynic might say that ableism marks the revenge of the disabled, since it would render normally abled people 'always already disabled'. In that sense, ableism should be counted as the great levelling ideology of our times. Perhaps unsurprisingly, the leading scholar-activist of ableism, Gregor Wolbring (2006), describes himself with pride on his website as 'a thalidomider and a wheelchair user'. From the sort of political economy that has consistently motivated the promotion of CT, ableism may be seen as a high-tech reproduction of the problem of 'positional goods' that afflicts modern welfare states, which is captured by the phrase, 'Keeping up with the Joneses' (Hirsch 1976). The problem arises when, for example, everyone in the labour market has already reached a certain level of competence but the means are available for those, if they wish, to boost their competence – say, through brain-boosting drugs. These people then shift the standard of competence upward, which in turn pressures would-be competitors to adopt a similar practice simply to remain competitive. Moreover, insofar as one selects on the basis of targeted competences instead of the whole person, there is an incentive to discount the downsides of, say, the brain-boosting drugs. Thus, the fact that, say, Modafinil might lower creativity as it increases alertness appears to be a price to all sides to the transaction are willing to pay, even though onlookers remain sceptical (Talbot 2009). In this respect, ableism is already in our midst.

Ableists also know enough about modern biology to realise that, left to its own devices, an accelerated version of natural selection is unlikely to result in creatures that we would be proud to call our own successor species. While evidence of common descent would no doubt remain in the genetic make-up and even the morphology of these later creatures, abilities valued in the earlier creatures might well have been eliminated because of intervening changes to the selection environment. Again, consider the relationship between extant reptiles and extinct dinosaurs: The mighty Tyrannosaurus would admit only with embarrassment its genetic responsibility for today's puny lizard. Without denying the fallibility of the science on which they rely, ableists prefer not to leave this matter purely to chance.

In other words, for a pro-enhancement policy not to appear Sisyphean, one must believe that Mendel trumps Darwin – that artificial selection *can* beat natural selection. A consequence of this belief is that one might continue to value the indefinite promotion of, say, cognitive ability but come to realise, given changes to the natural world, that cognitive ability is best conveyed by creatures that significantly differ from our own biological make-up but whose creation is nevertheless within the range of our technological powers. One might regard such 'enhancements' in ontologically modest terms so that our cognitively superior successors look like us, or at least share the same material substratum – that is, they are carbon-based. The prospects for horizontal gene transfer, which revisits the Lamarckian idea that our offspring might be decisively affected by physical changes in our own lifetimes, would likely prove a first step in that direction (Dyson 2007). For example, to enhance cognitive ability in an oxygen-deprived environment (assuming massive air pollution), the solution may be gene therapy based on some non-human species already able to get around this problem, from which then our offspring might also benefit. In this respect, ableism is a natural ally of the so-called adaptationist perspective on global climate change, which argues that rather than trying to deny or even stop climate change, the best course of action is to 'adapt', which may of course entail adapting our bodies as well as our external socio-economic systems (Stehr and von Storch 2005).

But of course, given more radical changes in the physical environment, the relevant sense of enhancement might move away from a carbon material substratum altogether to a more resilient silicon one that enables consciousness to be downloaded into computer androids. Put bluntly, Harris fails to see that a natural extension of his argument is a license to write us out of existence by disaggregating 'the human' into a set of capacities, each of which can be assessed and extended separately without the others that have been associated in evolutionary history with the human condition. Thus, the ableist aims to make good on an assertion that was originally treated as highly controversial when the UK bioethicist Jonathan Glover (1984) uttered it a quarter-century ago: 'Not just any aspect of present human nature is worth preserving.' Like many transhumanists, Harris conflates the 'superman' image of the transhuman (i.e. better humans) with the 'cyborg' image, which is a more likely outgrowth of CT-based enhancements: i.e. incorporation of hybrid carbon-silicon entities (including genetic xenotransplantation) that will likely reorient people's sentiments so as not to privilege the human. In the late 1980s, Donna Haraway (1991) promoted the cyborg image – then a staple of science fiction – as a model for feminism, given that 'human' meant white male humans. However, it's not clear whether female humans (black or white) benefit from this proposed redistribution of sentiment.

At the very least, under the ableist regime Harris countenances, the distinction between 'abled' and 'disabled' would be both relativised and modularised. This, in turn, would tend to expand the definition of 'disabled' from its traditional meaning (i.e. physical disability) to include a broader but vaguer category like 'disadvantaged' (aka 'non-competitive' or 'non-adaptive'), into which individuals may fall not because of any change to their bodies but, on the contrary, simply because their bodies fail to change in accordance with the norms of what Nikolas Rose calls 'somatic expertise'. Thus, people may come to think of themselves as 'always already disabled', that is, on the verge of falling behind in a social world where regular neurochemical upgradings are expected as a precondition for adequate performance. Whether this relativisation of disability actually benefits or simply marginalises even further those traditionally treated as physically disabled remains an open question.

The first public stirrings of the normative problems surrounding ableism concern the use of drugs to enhance competitive athletic and academic performance. The political responses so far suggest that this feature of the ableist agenda may well be subject to considerable regulation but it is very unlikely that its advance will be stopped altogether, notwithstanding one rearguard expression of consternation by the distinguished communitarian political philosopher Michael Sandel (2007). Sandel argues that ableist ideals violate the integrity of well-established social practices – including games – that rest on norms of fair play. Perhaps the most thoughtful discussion of this issue comes from a clinician at the University of Pennsylvania medical school, who attempts to draw lessons from the history of cosmetic surgery, which, after having begun as war-related reconstructive surgery, developed in a largely unregulated fashion in the consumer market (Chatterjee

2007). The author, a self-described 'cosmetic neurologist' who in the spirit of facelifts tightens synaptic connections for customers in search of that extra competitive edge, draws on Alfred Adler's classic 'inferiority complex' theory in converting cosmetic surgery into a free-floating biomedical treatment.

Why is Harris blind to such prospects for enhancement? Despite his progressive rhetoric, Harris shares with his opponents – who include not only Sandel and the Aristotelian Leon Kass (George W. Bush's bioethics tsar) but also Jürgen Habermas and Francis Fukuyama – a belief in an ontologically robust idea of human nature. But this idea is not borne out by either Darwin's own purely conventionalist account of species identity or the general drift of transhumanist thought towards a 'posthuman' condition. Indeed, Harris looks progressive only because of the primitive state of the most controversial enhancement technologies. This means he can have his cake and eat it: He can gesture towards a transhumanist future but for now his hardest cases concern the prospect of humans in more-or-less their current embodiment living indefinitely (Harris 2007: 67–8). To be sure, such cases raise interesting metaphysical questions, given the long-standing link that Western culture has forged between the meaning of life and the inevitability of death. However, it will not be long before advances in enhancement technologies broaden the metaphysical issues to include what the medieval scholastics called 'the problem of universals', namely, how can the same form be communicated in different configurations of matter. More concretely: How would one determine whether an entity substantially different in material composition from today's humans is still human – or at least sufficiently human to merit the value normally invested in humans?

At first glance, Harris' faux progressivism reflects the familiar philosopher's flight at dusk, to recall Hegel's line about the Owl of Minerva. In other words, *Enhancing Evolution* mainly provides reasons for discarding positions that the onward march of science has already made irrelevant. However, their irrelevance has yet to be fully appreciated because these 'undead' positions are conveyed by the likes of Habermas (2002) and Fukuyama (2002) who for now remain prominent in public intellectual life. For the most part Harris rightly rejects their views, though sometimes his arguments could be more forceful.

For example, Habermas worries that genetically designed offspring would lack any sense of moral autonomy by virtue of having been – and knowing to have been – produced as means for realising the ends of parents who, say, wanted a child with certain looks and talents. Harris counters by observing that child-rearing has been always to some extent instrumental, the only difference now being our enhanced capacities for strategic intervention: Matters that in the past were dealt with diffusely by, say, placing the child in a certain environment are increasingly treated in a more focussed fashion with drugs or even germ line manipulation (Harris 2007: 137–42). But this utilitarian response is unlikely to sway Habermas, for whom autonomy is non-negotiable at any price. Harris would have done better to stress that autonomy has been always a procedural, not a substantive, value. In other words, we respect people's autonomy by treating them a certain way, *regardless* of what we know about them. Thus lies the wisdom of John Rawls' (1971) 'veil of ignorance' as the original position from which to determine the fundamental principles of justice. But more importantly, the material basis for attributing autonomy may be strengthened by enhancement research, much of which aims to reverse the effects of prior causes, ranging from the use of stem cells in regenerative medicine to the removal of memory traces, as depicted in the 2004 Hollywood film, *The Eternal Sunshine of the Spotless Mind*. The result is to expand both the physical and the psychological sphere of action, overturning the commonsense view that age necessarily narrows our existential horizons.

But Harris' blindspot goes beyond his philosophical obsession with telling history's losers exactly why they have lost. He is almost completely blind to the truth contained in their concerns, perhaps because he is so lacking of a religious sensibility. The missing link between Hegel and Marx, Ludwig Feuerbach became notorious for arguing that the Judaeo-Christian God was simply the alienated projection of all that humans valued in themselves, only now used to judge and dominate them. To be sure, there are both empowering and disempowering features of this cognitive tendency. Feuerbach, himself a theologian by training, was debarred from the academy because he promoted Human-ism as an empowering religious successor to Christianity. Other post-religious practices have included state-worship and the identification with corporate entities more generally. While Hobbes' *Leviathan* and Hegel's *Philosophy of Right* may be read as relatively even-handed treatments of the pros and cons of such alienation, Marx, Freud and many 19th and 20th century thinkers have stressed the pathological dimensions. A radical transhumanist movement like ableism aims to redress the balance by justifying human self-sacrifice for the sake of some other being that more fully realises what we most value in ourselves. Not surprisingly, in the hands of gurus like Ray Kurzweil, research into artificial intelligence and artificial life looks like high-tech political theology, what the popular writer Erik Davis (1998) has called 'TechGnosis'.

I said that Harris is 'almost completely blind' to the radical nature of the transhumanist challenge. The one aspect he sees is the need for people to participate more actively in scientific research relating to enhancement, what he euphemistically urges as their 'mandatory contribution to public goods' (Harris 2007: 196). Harris justifies such participation, despite its risky aspects, on both scientific and moral grounds: Not only is it likely to improve the range and quality of the scientific findings but also it addresses our obligations to promote our own and future generations. Needless to say, were public participation in enhancement research to attain the status of jury duty, it might also establish good will for a form of inquiry that is bound to challenge our sense of who we are in the years to come.

7 Conclusions: The prospects for the CT agenda

One should not think of the disciplines involved in the CT agenda as somehow driven by their separate paradigms towards convergence, which once fully realised can then be applied for the benefit of society. On the contrary, the relevant sciences are pursuing many different agendas at once, progress in which is currently driven by the client base – not least its patience to wait for the relevant breakthroughs that would serve its interests. This state-of-affairs has rendered biology a financially successful but intellectually incoherent discipline, which philosophers sometimes dignify by saying that the science operates with a 'disunified ontology' (Dupré 1993). In the next chapter, I challenge this permissive philosophical attitude, which is strongly associated with the Neo-Darwinian paradigm. In any case, biology's equivocal research frontier is in full display for all to see: People who call themselves 'biologists' are driven, on the one hand, to search for 'deep' explanations for social traits already present in species that evolutionarily preceded it and, on the other, to reverse that implied history through micro-level manipulations of the sort associated with CT.

Under the circumstances, overlap in the client bases probably better explain any existing tendencies towards convergence than some philosophically inspired notion that independent lines of free inquiry tend to converge on a common truth. It is easy to imagine a public relations firm or advertising agency devoting its research division to both evolutionary psychology and CT research, the former for producing knowledge about what cannot be changed about human response patterns (which means indirect market strategies that play on those hard-wired biases) and the latter for knowledge of what can be changed (which may mean further investment in such changes so as to avoid the need for indirect marketing).

But of course, the private sector need not have all the fun. States and inter-state bodies – as long as they remain major players in the funding and regulation of scientific research – are in an unusually good position to provide direction at both the level of theory and application. A realistic starting point for policy is not a generalised scepticism towards the promised enhancement technologies associated with CT but an expectation that many will come to pass, albeit perhaps in diminished form. In any case, a minimal state or inter-state response would be to ensure that current socio-economic inequalities are not exacerbated by the introduction of enhancement technologies in a market environment. Of course, a more proactive policy would be preferred, especially one prepared to quickly incorporate enhancement technologies into established social welfare systems, while monitoring the consequences of mass adoption and restricting access outside those recognised systems. However, here two obstacles need to be overcome:

(1) There are principled objections from a broadly natural law standpoint about the violation of 'human being'. Rather than giving the religious origins of this concern a free pass, as a gesture to political tolerance, it will become increasingly important to contest the empirical basis for its concerns – Is everything about the human body sacrosanct? If so, why? These matters have been seriously contested within the theological traditions of Judaism, Christianity and Islam, and so there is no reason to think that the most vocal and perhaps stereotypical religiously inspired objectors to enhancement are representative of all considered opinion.
(2) However, a more substantial long-term problem is the element of risk that individuals will need to assume as new enhancement technologies are made generally available. The increasing concern with protecting human subjects during clinical trials and other experimental settings merely offloads the difficult question of the conditions under which a proposed enhancement is considered sufficiently safe to be made available *en masse*. It is unlikely that

there will ever be a clear answer. Indeed, there are likely to be major failures along the way, though hopefully not on the scale associated with faulty eugenics policies in the past. Nevertheless, states and inter-state bodies will need to provide some sort of welfare safety net or insurance against the risks that individuals will obviously undertake – and be encouraged to undertake – by subjecting themselves to enhancement regimes.

4
A Theology 2.0 for Humanity 2.0: Thinking Outside the Neo-Darwinian Box

Chapter 4 concerns theology's relevance to the future of humanity. Abrahamic theology is the original human science, in that the Bible is the first document that clearly defines humans as creatures 'in the image and likeness' of the world-creative deity. It follows that the distinctly human aspects of our being are those most closely oriented to God. This explains the perennial preoccupation with 'consciousness', the secular descendant of the soul, as the mark of the human. But this Abrahamic heritage also accounts for our fixation on science as a long-term collective quest for the ultimate truth about everything, which looks suspiciously like a secular version of Christianity's salvation narrative, especially when science is viewed as a political technology to install a 'heaven on earth'. Unfortunately, in its guise as theology's secular academic successor, philosophy has too often stressed theology's dogmatic and apologetic side – albeit now in defence of the scientific rather than the religious orthodoxy. The first section draws attention to the several aspects of this phenomenon. The remaining four sections shift the focus to a more empowering science-oriented theology – that is, *Theology 2.0* – suited to Humanity 2.0. Sections two and three deal with aspects of the philosophy and sociology of intelligent design theory, which re-instates the problem of divine creation at the heart of the scientific enterprise. Finally, sections four and five consider two historical (and heretical) paradigms for a Humanity 2.0 founded on a Theology 2.0: Joseph Priestley and Pierre Teilhard de Chardin.

1 Theology at its worst: Philosophy of science as neo-Darwinian apologetics

In the 20th century the role of philosophers *vis-à-vis* scientists, especially in the English-speaking world, shifted from being 'philosophers *of* science' to 'philosophers *for* science', a transition that is symbolised by the increasing significance given to Thomas Kuhn (Fuller 2000b; Fuller 2009a: chap. 2). In the case of Neo-Darwinism, the dominant paradigm in biology since the end of the Second World War, it has meant that philosophers have tolerated the theory's internal interpretive tensions by loosening their own criteria for a good scientific theory. This loosening of philosophical standards probably reflects the strong cultural standing of Neo-Darwinism. Intelligent design theory, in its quest to achieve intellectual respectability as an opponent to Neo-Darwinism, has somewhat mimicked its opponent by adopting a conception of 'intelligent designer' just as open as that of the Neo-Darwinist conception of 'evolution'. I argue that neither strategy works well, either epistemologically or politically.

When doing my PhD in history and philosophy of science at the University of Pittsburgh, now over a quarter-century ago, I always wondered why so many otherwise quite interesting and intelligent philosophers insisted on portraying themselves as 'underlabourers' for science. I first encountered the term in Jerry Fodor (1981), where he traced the view back to John Locke's self-understanding *vis-à-vis* Newton, which was then updated by the logical positivists *vis-à-vis* the early 20th century revolutions in physics and now further updated by philosophers like Fodor interested in the foundations of the (then) newly emerging field of cognitive science.

From the start, I found the idea of philosophers as underlabourers vaguely demeaning, since Locke's contribution to Newton was clearly one of public relations: He converted Newton's conceptually powerful but mathematically-based theory into a respectable ideology – 'Newtonianism' – that could be endorsed by the innumerate. However, with hindsight, I have come to believe that Fodor portrayed himself and his co-workers in cognitive science, as well as the positivists (and the Popperians, for that matter), in a needlessly unflattering light. While all of these philosophers were very interested in the foundations of the special sciences, they tended to propose theories and even methodologies that went against the grain of the scientists' self-understanding of their everyday practice. The fact that the philosophers no longer talked about metaphysics 'as such' did not mean they had lost their philosophical scruples.

If anything, their move from metaphysics as an autonomous discipline to metaphysics as scientific foundations was interpreted by many scientists as an aggressive move into their terrain.

It was the scientific backlash to such shamelessly critical philosophising that provided an audience for Michael Polanyi and Thomas Kuhn, two avowedly 'post-critical' philosophers who defended normal science as a philosophy-free zone. The sociology of scientific knowledge and its disciplinary successor, science and technology studies, have since that time followed Polanyi's and Kuhn's post-critical lead by presenting themselves in purely descriptive, not normative, terms (Fuller 2000b: chap. 7). Although each of these developments appeared for a certain time to be placing the philosopher and the scientist at a level playing field – what Yang (2008) has recently dubbed a 'fraternal' rather than a 'paternal' relationship – it is not obvious that such a status of equality is really tenable. Rather, philosophers and scientists seem destined to exist in some sort of relationship of subordination, with either one or the other party on top.

All of this is by way of introducing a group of contemporary philosophers who I take to be true spiritual heirs of Locke's underlabourers, namely, the *Neo-Darwinian apologists*. There are several representatives of this species – including David Hull, Michael Ruse, Elliott Sober and Daniel Dennett, not to mention younger variants. Their tone, style and emphasis may vary significantly but they all share warm feelings for Kuhn, in whose name they labour under a paradigm called the 'Neo-Darwinian synthesis'. The very name is significant for three reasons, each of which raises a host of interesting historical and philosophical questions:

(1) *'Neo-'*. There is no 'Neo-Newtonian' paradigm because for the 200 years following the publication of *Principia Mathematica,* physics fully exploited Newton's theoretical resources to try to resolve standing anomalies in his original account of the cosmos, especially relating in matters relating to light and energy. But then in the early 20th century, the discipline moved on to Einstein and beyond without returning to Newton for theoretical guidance (yet retaining a circumscribed version of his empirical achievements for practical purposes). However, prior to the emergence of experimental genetics as a research programme at the dawn of the 20th century, Darwin's theory of evolution of natural selection was widely taken to have already run its course in biology. At that point, Darwin was being kept afloat largely as a political ideology and a suggestive sociological

framework, what we now call 'Social Darwinism'. Thus, the phrase 'Neo-Darwinian' testifies to the role of Mendelian genetics in enabling Darwin's scientific resurrection: It finally provided an explanatory mechanism for natural selection, a process that had been previously understood only in terms of the shape of natural history that it allegedly produced. Nevertheless, we might still wonder about the exact point of grafting Darwin's original theory to a science, genetics, whose own research trajectory can be understood without any specific commitment to natural evolution, as it moved from a population to a molecular basis starting in the 1930s, which eventuated in the discovery of DNA as the genetic code and the routine sequencing of genomes (Morange 1998).

(2) *'Synthesis':* Of course, the Newtonian paradigm was itself a synthesis of disparate theories and phenomena, a point that William Whewell especially celebrated with such Latin coinages as 'colligation' and 'consilience', which laid the foundation for what is now called 'inference to the best explanation'. However, we do not normally refer to the 'Newtonian synthesis' because Newton and his successors removed the seams that originally divided the components of the synthesis, largely by homogenising the methods by which disparate physical phenomena were studied. Before Newton physical motions on Earth and in the heavens – including light and magnetism – had not been persuasively presented as subject to the same research programme because, under the influence of Aristotle, they were seen as possessing different natures and hence had to be studied differently. But after Newton all of these fields shared a common 'ideal of natural order' based on the regular movement of the planets in the solar system (Toulmin 1961). This contrasts with the case of the Neo-Darwinian synthesis. Here fundamental disagreements remain over which of the various constitutive disciplines should set the standard against which the contributions of the other disciplines are judged. For example, while both palaeontologists and molecular biologists call themselves 'evolutionists', their operational definitions of evolution differ markedly, with one field regarding as hypothetical (if not probably false) what the other field regards as established (if not incontrovertible), and vice versa. This particular disagreement was on clear public display in the final quarter of the 20th century, courtesy of Stephen Jay Gould and Richard Dawkins. In the wake of this dispute, scientific creationists and intelligent design theorists have capitalised on it for their own purposes (Woodward 2003), while the Neo-Darwinian apologists have taken it as an invitation

to settle the matter by philosophical means 'once and for all', sometimes quite explicitly (e.g. Sterelny 2001).

(3) *'Darwinian':* The expression 'Neo-Darwinian synthesis' is perhaps most clearly associated with Theodosius Dobzhansky (1937), who actually embodied the synthesis. Originally trained as a natural historian in Russia, he migrated to the US where he eventually succeeded his teacher T.H. Morgan at Columbia as head of the world's leading genetics laboratory prior to the revolution in molecular biology. Throughout the middle third of the 20th century, the term 'Darwinian' was used rhetorically to capture a sense of *natural* evolution that did not veer into the eugenically manipulative forms associated with other forms of evolution, notably the Soviet Union's revival of Lamarck's theory of inheritance of acquired traits. In morally abhorrent cases where there was a clear reliance on both Darwin and Mendel – notably Nazi racial hygiene – the phrase 'Social Darwinism' was extended to cover not only the treatment of natural selection as Adam Smith's invisible hand writ large in nature (a reading justified by Darwin's reliance on Malthus and the work of his own grandfather, Erasmus Darwin) but also deliberate policies of genocide, which Darwin himself clearly never advocated. (In fact, Darwin did not believe that our knowledge of heredity justified even the original eugenics proposals of his cousin, Francis Galton, to improve the species.) That 'Darwinism' could be modified by 'social' without the result appearing needlessly verbose spoke to Darwin's own caution in excluding any substantive discussion of humans from his landmark work, *On the Origin of Species*. This was especially helpful after World War II, when Darwin's name could be easily invoked – unlike, say, Herbert Spencer's – to defend the idea that natural selection applied only to prehistoric, not historic, time. This provided a politically correct way of dividing the work of biological and social scientists that remained intact until the mid-1970s, with the publication of E.O. Wilson's *Sociobiology* (1975) and Richard Dawkins' *Selfish Gene* (1976). At the same time, the conceptual independence of genetics research from the rest of the 'synthesis' has come to be re-visited with the revolution in molecular biology, which for the past half-century has been the most active area of biological research, increasingly biotechnology, where matters concerning the actual history of the Earth and the original formation of species are not especially relevant.

It is an interesting sociological fact that the scientists who would normally be regarded as the main empirical researchers in the 'Neo-Darwinian synthesis' do not especially resonate to that phrase themselves. If anything, they tend to regard the invocation of 'Darwinian' as a Creationist ploy to conjure up all sorts of unsavoury cultural associations – especially heartless capitalism and vicious Nazism – that detracts from focussing on the 'real science'. Thus, biologists much prefer the neutral expression, *modern evolutionary theory* or *modern evolutionary synthesis*. These expressions serve a dual function for the scientists: They remove any historical trace and they keep the future open as to what 'evolution' might come to mean (i.e. not simply or even primarily Darwinian mechanisms).

Unfortunately, to the ears of an underlabouring philosopher, 'Darwinian' remains important to keep in the phrase for two reasons. First is the *positive* cultural association of Darwin with secularism, naturalism and even ecology. Second is the potential unclarity, if not unfalsifiability, of biological theory if specific mechanisms are not identified as primary in the evolutionary process. These two reasons reveal that, even in their underlabouring capacity, philosophers are still fond of Popper's (1946) conception of science as an 'open society'. Thus, for them the term 'Darwinian' symbolises at once science's progressive yet self-critical character. Nevertheless, one must admit that the Neo-Darwinian apologists find themselves in a peculiar rhetorical position, given that those for whom they provide apologetics do not see the need for their services!

So, what are we to make of philosophy of biology's unrequited love of biological science? First, philosophers are more invested than scientists in the idea that the synthesis remains intact with Darwin's theory of evolution by natural selection functioning like Newton's Laws in the old 'covering-law' accounts of unified science favoured by the logical positivists. Thus, Daniel Dennett (1995) has literally applied William James' turn-of-the-century quip that natural selection acted as a 'universal solvent' to remove superstition from every belief system it touches, in order to convert Darwin's theory – originally intended as a generalisation about Earth's natural history – into an all-purpose model that might even explain how we happen to live in the particular physical universe that we do. However, this continuation of Newton-sized philosophical ambitions by Darwinian means raises many problems – or, more optimistically, provides many opportunities – for Darwin's apologists. In particular, the semantically relaxed conception of evolution favoured by practising biologists leads in many different directions, which then become the source of deep hermeneutical tensions for philosophers that are comparable to the problems faced by those collating the differing accounts of Christ's life given in the Gospels.

In light of the above discussion, consider the several different senses of 'evolution' that biologists routinely move between, depending on their particular research speciality and topic of investigation:

1. *Common descent with modification* – closely associated with Darwin himself and especially favoured by palaeontologists, though also easily contested by the presence of 'gaps' or 'leaps' in the fossil record, a point exploited by creationists and some intelligent design theorists – often aided by strongly antirealist evolutionary scientists, such as Stephen Jay Gould.
2. *Increasing differences in DNA* – the so-called molecular clock hypothesis, a version of (1) updated in light of molecular biology, which associates the differences in the genomes of two species with the number of mutations they have undergone since their ancestral populations divided from a common gene pool, which in turn enables inferences about the age of the species.
3. *Non-random change in the frequency distribution of traits in a gene pool* the classic Mendelian definition of evolution, but also consistent with so-called 'neutral evolution', whereby most genetic change turns out to be the product of 'drift', that is, a statistical by-product of natural selection. Another sense in which this view is 'neutral' is that it can be used to understand evolution as either a natural or an artificial process – as it were, before and after the God-like intervention of humans.
4. *Increasing control over nature* – a stronger version of (3) that presumes that humanity will render natural selection a completely artificial process as we take more control of the environment. The heyday for this view was the cybernetic revolution of the 1950s and 1960s, especially the work of Gregory Bateson, who equally warned of backlashes. A diminished version survives in Dawkins' concept that the technological infrastructure of modern life constitutes our 'extended phenotype'.
5. *Increasing complexity and adaptability* – an idea that Herbert Spencer carried over from Lamarck to Darwin, picked up again by Julian Huxley in the first book to use the phrase 'evolutionary synthesis', and which lives on in the writings of Richard Dawkins evolutionary psychologists. Contrary to Darwin's own rather principled, proto-Peter Singer, views about the fundamental equality of all species under the eyes of natural selection, this view hints at biological criteria for our species uniqueness, if not superiority.

6. *Convergent evolution* – a view that is outright counter-Darwinian in its suggestion that over time the possible forms of life narrows, indeed converging on increasingly similar forms that may involve the recurrence of atavisms – i.e. genetic throwbacks that effectively are recycled to produce new organisms or adaptations. Biologists who hold this view, such as Simon Conway Morris (2003), tend to be theists or Lamarckians, but in any case opposed to the Darwinian purism of Gould (1988), who argued that were evolutionary history replayed, a radically different array of species would result.

As of this writing, there is no agreed formulation of the Neo-Darwinian synthesis comparable to the deductive formulation of Newtonian world-system in its 19th century heyday. Instead, over the past quarter-century, philosophers of science have shifted their criteria of an adequate scientific theory from the Newtonian gold standard of a systematically unified, mathematically expressed account of nature to the much looser one, whereby a theory becomes no more than a collection of models, each of which provides a partial representation of nature's complexity that can together figure in a narrative account of evolution. In terms of philosophical homelands, one might call this the great shift from Vienna (in the 1930s) to Stanford (in the 1980s), as canonised in Galison and Stump (1996).

Perhaps the main – certainly most noticeable – challenger to the Neo-Darwinian synthesis today as an overall explanation for the nature of life and the origin of species is intelligent design theory, which proposes to treat nature as an artefact in a very robust and literal sense, namely, as implying the existence of an intelligence responsible for the design. This idea was fundamental to the Scientific Revolution's radical interpretation of the Biblical idea that humans are created 'in the image and likeness of God', which was read to imply that nature is God's machine, which we can understand by virtue of our own ability to make machines (Fuller 2007a: chap. 6). This view was also central to English natural theology, a hybrid of scientific and religious thought that flourished well into the 19th century. Its representatives included such figures as Joseph Priestley, William Paley, Thomas Malthus and William Whewell. These figures ranged over the entire political spectrum of the day, but they tended to treat nature as a single purpose-built functioning system that operates according to its own economic principles to make maximum use of the available energy. Indeed, these figures believed that a systematic vision of nature was required for the possibility of systematic scientific inquiry.

Kant famously began – and Darwin largely completed – the intellectual drive against natural theology by distinguishing the (strong) psychological compulsion behind its view of science from its (unproven) epistemological basis. Nearly two centuries later, intelligent design theory is now trying to reverse this Kant-Darwin move in thought, aided by a generation of theologically inspired scientists trained mainly not in Darwin's own field studies and natural history, but chemistry, engineering and applied statistics, often with a strong grounding in computer simulations. Intelligent design theory has run into many legal and political battles in the United States, whose limits on the expression of religion in publicly funded schools have been used against the theory by Neo-Darwinian apologists. For them, intelligent design theory is 'born again creationism'. One consequence has been that intelligent design theorists tend not to talk about the properties of the 'designing intelligence' behind nature, perhaps even implying that life could have been seeded from an extraterrestrial source, as was suggested originally by the great Swedish chemist Svante Arrhenius a century ago and updated by the co-discoverer of the double helix model of DNA, Francis Crick. In that respect, the theory's proponents have tried to treat the concept of 'intelligent design' very much as Neo-Darwinists have treated 'evolution', namely, as a 'big tent' for many different competing interpretations that do not necessarily add up to a coherent or compelling theory.

The failure of intelligent design theory to specify the intelligent designer constitutes both a rhetorical and an epistemological disadvantage. Neo-Darwinian opponents have derided theory as, in principle, allowing for a 'flying spaghetti monster' to count as a possible intelligent designer. The epistemological disadvantage is subtler, namely, that intelligent design theory is unnecessarily forced to adopt an instrumentalist philosophy of science, whereby its theory is treated merely as a device for explaining particular phenomena (i.e. as products of intelligent design) without allowing inferences to the best explanation (i.e. the properties of the implied designer). Meyer (2009) is a recent systematic attempt to inject a more scientific realist perspective into intelligent design theory, but he too stops short of introducing what I believe is a *necessary* return to theology as the source of theo-retical guidance on the nature of the intelligent designer (Fuller 2008a).

By way of conclusion, to make this point, consider Elliott Sober's recent forensic investigation of the epistemological warrant for both Neo-Darwinism and intelligent design theory, *Evidence and Evolution*

(2008). Two of his main arguments against intelligent design theory may be obviated if the theory was more open about its theological commitments. I list them below:

1. *Intelligent design theory invents assumptions on an ad hoc basis to explain the allegedly designed character of aspects of nature, such as the panda's thumb, that most probably did not arise by design.* Sober's argument works as long as there is no theory of how the designer designs, namely, the principles behind the deity's handicraft governing different levels of nature, say, comparable to how we infer the architectural principles underlying an ancient edifice. This would involve imputing to the deity a psychology of sorts, one akin to Herbert Simon's (1977) 'bounded rationality', which portrays the rational agent as a constrained optimiser, that is, someone who works toward the best possible overall outcome, which in turn may require the tolerance of suboptimal outcomes along the way. This mode of thinking was common in the late 17th and early 18th century heyday of theodicy, the branch of theology concerned with justifying the horrors of nature and evils of humans in a world supposedly created by an omnibenevolent and omnipotent God. However, theodicy always had a borderline heretical status because it presupposes that humans can second-guess God's motives.
2. *Even if intelligent design theory were correct that every event must have a cause, and every species must have an intelligence behind its design, it does not follow that the cause or the intelligence need be the same in all cases.* Sober's argument here cuts very deep – perhaps even too deep for Sober himself, since it potentially undercuts the idea that there is an intelligible unity to nature that provides science with its goal and guiding impulse. In this respect, Humean scepticism towards the cosmological argument for the existence of God, whereby all causal chains are traced back indefinitely with no convergence at an ultimate source, also cuts against the point of Newton's project of unifying the diversity of nature under the fewest number of laws: Why engage in Newton's project at all, if there is not a single source to all things? (In this respect, 'big bang' cosmology, the product of the 20th century Jesuit natural philosopher Georges Lemaître, might be seen as trying to bridge the gap between Hume and Newton, especially if the origin of the universe is seen as arbitrary.) However, this scepticism could be mitigated, if not completely overturned, if additional theological arguments were presented that favour monotheism over the sort of polytheism that is con-

sistent with the Hume but would have unlikely issued in Newton's science.

In short, by studiously avoiding the appeal to theological arguments as part of their scientific explanations, intelligent design theorists only inhibit their own ability to meet the opposition of Neo-Darwinian apologists like Sober. Admittedly, making such appeals would mean not only re-opening old theological debates but also making them part of secular academic debate. A test of our collective intellectual maturity will lie in our ability to tolerate such a newly charged situation. But as it stands, intelligent design theory does itself no intellectual favours by keeping the identity of the intelligent designer as vague as Neo-Darwinians keep the identity of evolution, even if that practice appears justified as politically expedient. As someone interested to reintroducing debates about our relationship to the deity into the public sphere, not least in order to define more clearly who we think we are when we call ourselves 'human', I shall now explore the range of issues opened by this prospect based on my own participant-observation.

2 Theology at its best: Intelligent design as heuristic for scientific discovery

Future historians will have a field day trying to make sense of the fear, loathing and anger that have been generated over the past quarter century in the various controversies surrounding evolution, divine creation and, most recently, intelligent design as competing accounts for the origins and maintenance of life on earth. As of this writing, these controversies have migrated far beyond their genesis in the Christian American heartland to encompass the world, now sporting even Muslim and Jewish variants. At the public policy level, the animus fuelling the controversies is easily understood: Virtually all the cases involve decisions about what to include in the science curriculum in state-run high schools. Put baldly, what may count as authorised knowledge in official educational settings? The suggestion is that if students are not exposed to one or more of these views, their intellectual capacity to function as citizens will be seriously compromised.

What is immediately apparent about this framing of the problem is that, on the whole, it is *not* the one that philosophers normally use to decide matters of scientific theory choice. Philosophers (still) tend to operate with a much more backward-looking and static view of the task. In other words, they presume that competing scientific theories are in a race that ends at the moment a philosopher is taking the decision, in which case the theories' track records are used to justify their respective fates. In contrast, a school board – or a court considering an appeal to a school board decision – is focussed on what the *next* generation should learn in order to deal effectively with the life challenges they are likely to face. In that case, the theories' track records are read for clues to future developments. Nevertheless, it is easy for this prospective judgement to be confused with the philosopher's retrospective one due to a shared problem-framing device: *Both presuppose that a forced choice must be made between mutually exclusive alternatives.* I have critiqued this presupposition from the earliest days of my social epistemology, partly influenced by Larry Laudan's coinage of 'context of pursuit' in contradistinction to 'context of justification' as a basis for scientific theory choice (Fuller 1993: esp. chap. 4; Laudan 1981a).

In any case, philosophers and keepers of the public sphere produce their common problem-framing device differently. On the one hand, philosophers sharpen the differences between competing theories that might otherwise overlap in the relevant scientists' minds through feats of definition and deduction that result in some key outcomes of one theory logically excluding those of another. On the other hand, educators and lawyers fixate on the scarcity of textbook space and classroom time as the basis for presuming that any attention given to one theory *ipso facto* takes away from that given to another, which becomes problematic against an implicit assumption of what constitutes a minimum level of attention needed to present a theory adequately. In this respect, a measure of the sensitive nature of today's evolution controversies is that in *Kitzmiller vs. Dover Area School District* (2005), the US Circuit Court case where I served as an expert witness, the bone of contention was the requirement that science teachers read a statement informing students that intelligent design theory exists as an alternative to Neo-Darwinism, while acknowledging the latter's dominant position in contemporary biology. The teachers were not forced to teach intelligent design, and certainly not prohibited from teaching evolution.

The artificial character of both the philosophical and the political problem-framing devices should be obvious. Nevertheless, there are reasons for promoting the artifice. In the case of the anti-evolutionists, the reasons relate to their increasing self-description as 'anti-Darwinists', mindful of the significant philosophical differences amongst various theories of evolution. In particular, Darwinian natural selection purports to be a universal process that affects all forms of life equally. Darwin believed not only that our physical morphology and mental dispositions bear witness to our evolutionary past, but also that our knowledge of such matters, albeit unique to *Homo sapiens*, does not necessarily confer on our species any long-term survival advantage. Unlike such architects of the so-called Neo-Darwinian synthesis from the 1930s and 1940s as Theodosius Dobzhansky and Julian Huxley, Darwin himself did not believe that humans could ever take sufficient control of the evolutionary process to reduce natural selection to a version of the breeding practices of artificial selection. (This had been the context in which Darwin's contemporary, Gregor Mendel, discovered the laws of heredity.) More fundamentally, Darwin did not believe that evolution itself is driven by an intelligence with which human intelligence has an elective affinity, let alone by virtue of our intelligence having been created in a deity's 'image of likeness'. Here Darwin broke fundamentally with the original evolutionary theorist, Jean-Baptiste Lamarck (who still believed in God but never managed to persuade his contemporaries that the deity might create by treating life-forms as loaded dice), not to mention the dissenting brand of Christianity in which he himself was raised (Desmond and Moore 2009).

Much of Darwin's pessimism is traceable to an overarching metaphysical indeterminism whose roots are to be found in Epicureanism, a species of naturalism that arose in the wake of the Alexandrian conquest of Athens, which treats stable order as a transient illusion, unnecessary attachment to which is best given over to therapy. From this standpoint, nature's one constant is the ultimate disorganised form of being we call 'death'. Thus, Darwin probably found it grotesque that the leading intelligent design theorist of his day, William Paley, followed fellow Anglican minister Thomas Malthus in alleging that long-term positive benefits were to be had in the struggle for subsistence in increasingly populous human societies – namely, the identification of the 'fitter' species members. However, some of Darwin's pessimism about humanity's fate was specific to the human brain's poor design, which allows enlightened scientific ideas to co-exist with rather perverse and self-destructive fixed ideas, the latter being genetic holdovers from a primitive past. In this regard, Darwin repeatedly refused all manner of typically Victorian solicitations – including from his cousin Francis Galton of eugenics fame – to endorse science-based schemes to promote the human condition, usually on grounds of the greater misery that Darwin believed might ensue for those who are supposed to be helped, as well as the risks to which the lives of animals might be exposed (e.g. by their use in medical experiments). Amongst such dangerous schemes, Darwin even included John Stuart Mill's call for easily available contraception advice to prevent unwanted births (Peart and Levy 2008).

For nearly all creationists and many intelligent design supporters, the legally relevant question here is whether restricting publicly funded science instruction to the pronounced anti-humanism of Darwin's theory of biological evolution constitutes an encroachment of the state into matters that are constitutionally delegated to civil society. Thus, the legal mind behind intelligent design theory, Philip Johnson (1991), has accused the singular promotion of Darwinism in schools of fostering a naturalistic *religion*. And he is literally correct, as long as US courts insist on upholding the idea that science requires a belief that natural history is entirely the result of processes observable today under normal circumstances. This insistence ties 'naturalism' to what in the 19th century was called 'uniformitarianism' but which nowadays might be regarded as a species of 'inductivism'. It harks back to David Hume's rather muted defence of Newtonian mechanics as a mathematically elegant and useful summary of the solar system's regularities – but not a glimpse into the deep causal structure of the natural world – and hence not a basis for launching a design-based argument for God's existence.

Yet, it is also clear that the two most recent revolutions in physics – relativity theory and quantum mechanics – would have been precluded by such 'naturalistic' strictures, had they prevailed in the first quarter of the 20th century, when the epistemic status of the two theories was still very much contested. In both cases, there was general agreement over the relevant mathematics and experimental outcomes. The difficulty was trying to find a physical interpretation that did not entail the suspension of key Newtonian assumptions, which for the previous two centuries had been taken as synonymous with the physics of the everyday world. It is to the great credit of the physics community that by 1930 they embraced challenging conceptions of space, time and cause that broke decisively with what had passed for the so-called 'natural attitude'. To be sure, 'common sense' and 'ordinary language' conceptions of space and time have persisted in philosophy into living memory (e.g. Gale 1967), attempting to show that our normal way of relating to the world presumes the irrelevance if not outright incoherence of technical scientific concepts. Although the US National Academy of Sciences would be the last to see itself in these terms, its rearguard appeal to 'methodological naturalism' as its official ideology continues this embarrassing modern tendency for (some) philosophers to promote the scientific beliefs they inherited as if they were timeless epistemological truths (e.g. Pennock 2010).

In this context, it is easy to appreciate the appeal that the logical positivists and the Popperians – as well as the early Richard Rorty (1965) – made to cutting-edge science as doing a better job than professional philosophy in providing a Kant-style 'anticipation of experience' that defines philosophy at its best. Today's so-called methodological naturalists would be at a loss to deal with such figures. Common to the positivists, the Popperians and Rorty was a radical reading of Kant, in which our concepts riskily stake out the scope of all that there is to know, the empirical details of which would be then filled out by science. Theories are thus to be understood as hypotheses rather than dogmas, which in turn marks the break point between organised science and organised religion. As Popper (1963) especially stressed, science (done well) is simply (well done) philosophy but by technologically enhanced means. From this standpoint, the error committed by most 19th century forms of Kantianism – including Comtean positivism – was that they presumed that, in the wake of Newton, the pursuit of knowledge could proceed on *less* risky foundations than it had in the past. In effect, Kant's 'transcendental deduction' was interpreted as converting Newton's physics into metaphysics, which only served to reinforce both philosophy's and science's most conservative tendencies. These philosophers had failed to appreciate a point brought out very well by Bayes Theorem – namely, that it is possible for many knowledge claims to be empirically well-founded if certain theoretical assumptions are true while there remains a strong chance that those assumptions are themselves false.

This historical detour returns to my own interest in promoting intelligent design in schools, which is much more positive than Johnson's original worries about naturalism turning into an established religion. I actually believe that the deep theological roots of intelligent design theory provide a robust basis for perpetuating the radical spirit of inquiry that marks both philosophy and science at their best – not at their worst, as their collective response to intelligent design has put on public display (Fuller 2009b). As a true social constructivist (Fuller 2000b: Preface), I see myself as one of the constructors of intelligent design theory. I am not simply remarking from the sidelines about what others have done or are doing, as a historian or a journalist might. Rather I am making a front-line contribution to defining the theory's identity. There are many ways to be involved in such a task. Most obviously one could announce a new scientific finding on behalf of the theory – and perhaps be subject to various forms of censorship and ostracism in the process, or simply be told that there is already a Neo-Darwinian account for the phenomenon. But equally important is to reclaim earlier findings for intelligent design that Neo-Darwinism has illegitimately claimed for itself simply because of the 'winner-takes-all' nature of scientific paradigms, a process that Kuhn legitimated. The fact that even today most work in the disciplines that comprise the Neo-Darwinian synthesis – from palaeontology and ecology to genetics and molecular biology – can be conducted with only the vaguest of references to evolution is a reminder that those on whose shoulders the likes of Richard Dawkins would stand today have operated from a variety of explanatory frameworks, including ones that mix creationist and evolutionary accounts, as well as purely special creationist ones (e.g. Linnaeus, Cuvier, Mendel).

Although several evolutionists have won the Nobel Prize, they were for achievements that stand on their own regardless of the validity of Darwin's theory. This even applies to Dawkins' Oxford mentor, Nikko Tinbergen, whose empirical work in ethology is compatible with a variety of naturalistic explanatory frameworks. Here the contrast with physics is interesting, since some of its Nobel Prizes have been awarded for contributions to unified field theory, which is to say, something of comparable generality to evolutionary theory in biology. If nothing else, Nobel Prize committees are consensualist. Not surprisingly, the Crafoord Prize has been set up specifically to recognise contributions specific to evolutionary theory.

Indeed, the seemingly indomitable character of evolutionary theory in biology is based on a threefold illusion, which will inform the discussion that follows: (1) *Semantically*, 'evolution' is not a univocal term but one whose meaning can be widened (as in 'theistic evolution', 'directed evolution' or even 'creative evolution') or narrowed (as in 'Darwinian evolution'), depending on the argumentative context. (2) *Epistemically*, the bodies of knowledge used to construct the Neo-Darwinian synthesis can also be used – properly re-interpreted and supplemented by other bodies of knowledge (e.g. engineering) – to form other biological syntheses, including ones that foreground intelligent design. (3) *Historically*, many, if not most, of the knowledge claims that are now justified 'naturalistically' began life as 'supernatural' hypotheses. It would not be unreasonable to think of 'scientific naturalism' as a retrospective honour bestowed on supernatural hypotheses that have managed over time to command the assent of non-believers.

A good example of (3) is Newton's treatment of motion as something internal, not external, to matter (aka inertia), which in turn is subject to a principle of constant universal attraction (aka gravity), both of which adhere to strict mathematical laws. This was the physics of spiritual bodies guided by a higher intelligence. It was radically different from the classic Epicurean view of what Hobbes still called the 'state of nature' as successive moments of temporary configurations of material atoms forever in violent motion. While the latter may inspire – as it did in Hobbes – a metaphysics of power and control, only the former offers the prospect of a long-term intellectual project like science that rises decisively above the exigencies of survival. And even if it is true that all supernaturally motivated scientific insights are eventually absorbed into the naturalistic worldview, it does not follow either that the supernaturalism was unnecessary or that naturalism is the final word. Indeed, the secular philosophical position nowadays known as 'scientific realism' retains the supernatural impulse in the form of science's persistent drive towards self-transcendence, as it refuses to rest on its empirical and practical laurels, but rather strives to arrive at an ultimately unified account of reality.

In terms of pedagogical implications, my support of intelligent design goes beyond merely requiring that students learn the history and philosophy of science alongside their normal studies. It involves re-engineering the science curriculum so that its history and philosophy falls within its normal remit. However, recalling point (3) above, the devil lies in the detail of the textbooks and curriculum that would claim to make good on this channelling of the supernatural for scientific purposes. On this basis, I did *not* endorse the *Of Pandas and People* textbook on offer in the Dover Area School District as fit for purpose. But in contrast, consider a conversation that Linnaeus allegedly had in which he expressed dissatisfaction with a Lapp herdsman's direct appeal to God to explain the design of the reindeer's hoof (Täljedal 2010). This was not because Linnaeus disdained design explanations but because he believed that God's signature is most clearly found, not in the detail construction of organisms, but in what he regarded as the hierarchical character of the entire ecosystem, which humans are Biblically empowered to cultivate. Basically Linnaeus identified the increased productivity of nature – through ever efficient forms of resource management – with the realisation of divine goodness (Koerner 1999).

In the United Kingdom, which does not constitutionally separate church and state, the largely public-funded Christian high schools have been receiving and, to some extent, implementing curricular guidance along the lines I now advocate for more than a decade (cf. Jones 1998). The result is that there is no marked difference between the performance of students in year-end science examinations between Christian and secular schools. Indeed, in some institutions, the Christian students performed significantly better. This includes more fundamentalist schools where students were taught Darwinist principles for purposes of passing exams but not as a mandatory belief system to be incorporated into whatever research or policy to which they might someday contribute. An indirect measure of the ideological character of evolutionary theory today is the soreness that Darwinists register about this last point, even though philosophers down through the ages have frequently recommended a hypothetical attitude towards scientific theories. Indeed, a theory cannot be fairly tested unless a clear operational distinction is drawn between knowing its content and believing its truth (cf. Fuller 2008a: chap. 1).

Here we move into what may be the most controversial aspect of my position, namely, that the active promotion of a certain broadly Abrahamic theological perspective is necessary to motivate students to undertake lives in science and to support those who decide to do so. In important respects, this cuts against Johnson's original animus for anathematising naturalism as creeping religion. Taken at his word, Johnson wants to return to the US founding fathers' understanding of 'separation of church and state', which was invoked to prevent the establishment of a national religion but without denying religion's role in motivating inquiry and other aspects of human development. For Europeans especially, it is important to recall the highly devolved nature of educational authority mandated by the US Constitution: Individual states (not the federal government) may set standards of achievement, but exactly how they are met – in terms of textbooks and curriculum – is left to the taxpayers in the respective school districts. Court cases arise only when school instruction appears to contradict what citizens thought they were paying for. Thus, most of the plaintiffs listed in *Kitzmiller vs. Dover Area School District* were upwardly mobile liberal parents who reflected Dover's transformation from a rural backwater to a bedroom community for the Pennsylvania state capital. By the same token, we should not be surprised if in the next few years Johnson's worst fears are vindicated by a major lawsuit brought against some science instructor whose overzealous naturalism leads him to deny divine causation in a public school district whose tax base is funded mainly by religious believers.

To appreciate the profound sense of devolutionism in American education, consider the history of the famed Scholastic Aptitude Test (SAT), the results of which are used by most US universities to set student entry standards. The basic test, mimicking an IQ test in its coverage of verbal and mathematical reasoning, was developed in the early 20th century by a private psychometric firm. By mid-century it had become a convention through mass institutional adoption. Over the years changes to the SAT's form and content have been determined by a combination of feedback from test outcomes and specific court cases (e.g. concerning cultural bias). From a European standpoint, this has been a remarkably bottom-up process in which the federal and state governments have been conspicuously passive. However, such devolutionism should be understood as the residue of the US founding fathers' interest in promoting a non-denominational but recognisably Abrahamic civic religion that combined a strong sense of individual autonomy with the idea of mutually binding peer-based authority.

Had agreement been reached on the principles of an American civic religion, it would have probably looked like a version of the 'religion of humanity' that came to be championed by the French republicans, who themselves had been inspired by such critical readers of the Bible influential in the American Revolution as Joseph Priestley, Thomas Paine and, indeed, Franklin and Jefferson. (The sensibility remains enshrined in France's answer to Westminster Abbey, the Panthéon, which looks suspiciously like the US Capitol.) However, differences in the inherited styles of Christian worship probably explain why the civic religion idea took hold in Catholic France but not in non-conformist Protestant America. In any case, the persistence of appeals to Biblical authority and divine providence in US political rhetoric is difficult to ignore, not least in the case of Abraham Lincoln, whose willingness to provoke a risky civil war was defended on these terms, even as he studiously avoided church membership and evaded questions of personal faith. Whereas some have read Lincoln cynically, I take him to have been a sincere devotee of that elusive American civic religion.

The key Abrahamic residue is the idea that humans are privileged above all other creatures in their capacity to understand and control their place in nature by virtue of having been created in the image and likeness of the creative deity. The residue is one that the Enlightenment retains from Protestantism after having purged latter's adherence to the Bible of its 'corrupt' and 'superstitious' elements. Indeed, the very terms 'Abrahamic' and 'monotheistic' to describe the three religions that descend from the Hebrew Scriptures – Judaism, Christianity and Islam – were Enlightenment neologisms coined in the spirit of trying to capture their common essential truths, which in the future would be taken forward by self-legislating rational beings (Masuzawa 2005). The republican revolt against hereditary monarchy and the scientific revolt against clerical authority were the two main fronts on which this movement proceeded in the 18th and 19th centuries. However, the 'movement', such as it was, was never unified either politically or scientifically, a point conjured up by the names of Comte and Popper – two figures who took their common inheritance in rather opposing directions. Nevertheless, my view is that despite this divergence of paths, the significance that science continues to exert on humanity's self-understanding, despite all its failures and disasters, testifies to our lingering sense of theologically-based ontological privilege. Indeed, a thoroughgoing atheist who looked at science's overall track record, especially in the 20th century, and was concerned simply with matters of the sustainability of the natural world, would call for a radical scale-back of risky science-driven commitments that would reshape the planet to humanity's convenience (Fuller 2010: chap. 6).

There is still something to the Enlightenment idea of a civic religion in which science plays a central role as the site for a systematically organised critical rationality. But what remains specifically 'religious' about this 'civic religion'? Two aspects: (1) Science's findings are framed in terms of the larger significance of things, nature's 'intelligent design', if you will. (2) Science's pursuit requires a particular species of faith – namely, perseverance in the face of adversity – given science's rather contestable balance sheet in registering goods and harms (Fuller 2010: chap. 1). If, in contrast to these two 'religious' senses, we were to value science solely for its practical consequences, then support for science would likely disperse to a variety of client-centred markets, after a while losing all pretence of trying to achieve the ultimately true and comprehensive picture of reality, as the various markets generate niche ontologies and epistemologies tailored to specific client needs.

Here it is worth recalling that a principle routinely invoked to demonstrate science's commitment to explaining the most phenomena by the fewest laws, Ockham's Razor, was originally applied to show that the language-world relationship is more systematic – and indeed unified – than simply a set of pairwise correspondences between words and things, each explained by a specific mediating concept. In this respect, Ockham's Razor approximates what the 17th century rationalists called 'pre-established harmony', which was treated as a second-order feature of the world that required the existence of God in a way that the explanation of first-order phenomena did not. In other words, Ockham's Razor was meant mainly to distinguish the explanation of something in isolation from something as part of a larger whole, which taken to the limit requires the postulation of God. This in turn justified a unified approach to the pursuit of science as the search for overarching natural laws, as opposed to the mapping of irreducibly separate domains of empirical reality that one finds in, say, ancient Greece and Rome, India and even China – all of which produced sophisticated forms of inquiry yet outside the Abrahamic sphere (Fuller 1997: chap. 7). In this sense, I remain a 'Eurocentrist' – though I prefer the term 'Occidentalist'. It is very unlikely that science would have taken the course it has – and valued as much as it has been – were it not for the Abrahamic belief that humans were created in the image of God.

However, that historical accident does not by itself speak against its potential universal application. The philosophical distinction between the contexts of *discovery* and *justification* in science may be understood as born of this insight. After all, one would not insist on such a strict distinction in contexts, had one not imagined the idiosyncratic origins of ideas of universal import, where the paradigm case is Jesus. Perhaps the most thorough and most sympathetic historian of Chinese civilisation, Joseph Needham, anticipated my conclusion upon considering the uniqueness of the West's 'scientific revolution' (Needham 1976; cf. Fuller 1997: chap. 5). According to Needham (in a part of his argument that tends *not* to be challenged), the Abrahamic self-privileging of Europeans emboldened them to remake (aka rationalise) the entire globe, typically through science-based military innovations that forced the reconfiguration of trading patterns to their advantage, resulting in what Marxists call the 'international division of labour' that characterises capitalism as a 'world-system'. In this respect, there is no denying that the universalisation of 'Eurocentric' science is an outgrowth of Western imperialism, and there is a real debate to be had about whether the resulting benefits outweigh the costs.

After all, reductionism – the signature theory-driven project of modern science – makes sense only if all of reality is presumed to have been constructed by, so to speak, a single hand that happens to resemble our own. In that case, science becomes an exercise in reverse engineering but this time with an eye towards improving or completing what God – as opposed to some other human – had designed. In this story atomism plays a central yet circumscribed role. The difference between 'atoms' as discussed in ancient and modern atomism is that in the latter case atomic indeterminacy is conceptualised as 'structured' in a way – captured by probability theory – that overcomes fatalism and encourages scientific inquiry, thus enabling what Ian Hacking (1990) has called 'the taming of chance'. Little surprise, then, that reductionism's crisis in light of the irreducibly indeterminate nature of quantum phenomena was widely interpreted as marking a limit to humanity's divine pretensions. Nevertheless, our godlike ambitions were subsequently rechannelled into molecular biology and nowadays inhabit the promissory science of nanotechnology, which we have considered in Chapter 3 in the context of the 'converging technologies' agenda.

3 Theology as a source of empowerment in science education

Science education is not only about instilling a certain way of knowing but also of being – namely, as creatures whose knowledge enables us to take greater responsibility for what happens in the empirical world, as we better understand its *modus operandi*. (I have in mind the prospect that we might someday be capable of travelling to – and hence intervening in – the affairs of other inhabited planets, perhaps in some other galaxy.) It was confidence in achieving such a perspective that inspired the US founding fathers to draft a constitution that, through a complex system of separation of powers and checks and balances, would (where possible) anticipate and (if not) correct obstructions to good government, very much in the manner they imagined God to have created Newton's universe (Cohen 1995). Whereas Comte might be guilty of transferring the ritualistic trappings of the Catholic Church to his scientific utopia, Franklin and Jefferson – prodded by Joseph Priestley – potentially stood accused of reducing the exercise of Christian faith to the appliance of science. The legacy of this charge is the intensity of the legal disputes relating to the science-religion interface: Are the parties to such cases trying to relieve science of religious superstition or realise religion through scientific inquiry – or both? My own answer, with which I believe the founding fathers would agree, is *both*.

Moreover, this is by no means a uniquely American phenomenon. Both Thomas Huxley and his grandson Julian were concerned with how the spread of Darwinism might affect people's self-understanding and the value they invested in the pursuit of science itself. Whereas Thomas saw humanity's civilising professions – the law, medicine and engineering – as operating largely in defiance of natural selection, Julian, inspired by 50 years of genetics research, envisaged a 'transhumanist' condition whereby humans would actually take control of natural selection. But in either case, science had a moral obligation to promote a pro-human perspective (Fuller 2007b: chap. 5).

Once again let me recall the distinction between the contexts of *discovery* and *justification* in scientific inquiry – now, in the way I appealed to it in *Kitzmiller*, based on previous work and continuing into more recent work on the science-religion interface (Fuller 2000b: chap. 1; Fuller 2007b: chap. 4). This appeal cuts against the post-Kuhnian tendency to deny or blur the distinction by arguing either that scientists can discover only that which is justifiable or that scientific justification consists in an idealised discovery procedure. A sharp distinction is especially useful in discussing the possible role of religion in the science classroom. On the one hand, an Abrahamic framework has served as a powerful heuristic for conceptualising deep causal structures by challenging scientists (e.g. Newton, Faraday, Mendel) to operationalise divine agency in terms of experimental manipulations. On the other hand, when such operationalisations have resulted in robust empirical findings, they have been adopted even by scientists lacking the discoverer's original theistic framework. Taken together, these two points about, respectively, discovery and justification imply that religiously motivated inquirers can produce science worthy of acceptance by those who do not share their beliefs. This is a strong intellectual argument for the explicit discussion of religion in the science classroom – and indeed for encouraging religiously motivated students (at least in the Abrahamic tradition) to enter scientific careers.

What is *not* licensed is the advocacy of an unconditional religious commitment on the basis of whatever scientific successes resulted from applying a religious belief. However, a grey area is suggested by the very term 'Abrahamic', which as noted above is an Enlightenment coinage designed to locate the rational core common to the three great 'Religions of the Book' that promoted the advancement of science. The suggestion is that, once identified, this core might lay the foundations of a civic religion that could provide a reliable basis for scientific discovery, which in turn would serve to persuade those of other faiths and metaphysical commitments to convert to the civic religion. Starting perhaps with Burtt (1925), historians and philosophers of science have masked the controversial theological animus of this argument – science in the service of proselytism – by a euphemistic appeal to 'metaphysics' as an ineradicable feature of scientific inquiry. This is the context to which historian Herbert Butterfield's popularisation of a 17th century 'Scientific Revolution' – at first in a set of BBC Radio lectures – belongs (Butterfield 1949). He believed that it was both comparable to and continuous with Christianity as a world-transformative movement, as did William Whewell, who crystallised Newton's reputation as the greatest scientist by ranking him just after Jesus in significant humans. Regardless of whether one believes that certain metaphysical commitments are necessary for the conduct of science, our default sanguine views about science (which I share) presuppose an understanding of its place in the world that goes well beyond what is warranted solely by science's rather chequered track record (Fuller 2010).

Mindful of the above, Ian Jarvie (2010), a student of Karl Popper and founder of the journal *Philosophy of the Social Sciences*, is not unreasonably worried that science itself might become a kind of religion (aka scientism), which had been one of the targets of logical positivist suspicions about 'metaphysics'. I myself canvassed the religiosity of contemporary science in Fuller (2006a: chaps. 4–6), which largely casts the 'Science Wars' of the 1990s as a Protestant-style struggle staged by science studies scholars and other postmodernists against authoritative readings of their shared scientific heritage (i.e. *not* anti- or even pseudoscience). Jarvie argues that science and religion should remain separate, in part, because of religion's historically documented deleterious effects, many of which are traceable to its demand for unconditional belief, an epistemic attitude that is contrary to the spirit of science.

In response, I would first distinguish between first-order scientific theories, which I agree should always be given hypothetical status, and the second-order idea of science as a theoretical project, which may require specific (Monotheistic) religious attitudes. This is because there is more to science than simply a critical, open-minded and rational attitude toward the world. There is also the 'belief', if you will, that a series of errors might bring you closer to the truth, and so the path of inquiry is always worthy of one's perseverance upon which future generations might build. It is this belief that elevates science above a mere hobby, and indeed accounts for the unprecedented concentration of material and intellectual resources that, *pace* Jarvie, has arguably enabled science to do much more damage than religion in its briefer history, if the two are compared in the same sense, namely, as organised activities. (If one imagines religion in terms of its institutional embodiment but science as pure intellectual content, then of course in that question-begging sense religion looks worse than science.) However, this observation is not grounds to shut down science but to recall the doctrine of dirty hands: With the capacity to do much good comes the capacity to do much harm, and the acceptable ratio of costs to benefits is ultimately subject to negotiation and regulation. Moreover, if we consider both science and religion as organised activities, then it is useful to recall the roots of Popper's open/closed society distinction in Henri Bergson's distinction between open and closed religions, as in Protestantism and Catholicism or Buddhism and Hinduism, which in terms of science corresponds to Popper and Kuhn (Fuller 2003:

chap. 10). In short, Jarvie's distinction looks compelling because he is comparing *open science* and *closed religion*.

Another of Popper's students, Jeremy Shearmur (2010), raises a series of questions concerning my positive views of intelligent design as a scientific research programme, especially in terms of how they might differ from the sheer anti-Darwinism and anti-naturalism that appears to dominate Philip Johnson's agenda. First, I would say that a careful reading of the various historical and contemporary theorists of intelligent design reveals a diversity of opinion about the identity – or even the identifiability – of the intelligence informing nature's design comparable to the diversity of processes endorsed by self-avowed 'evolutionists'. It is unfortunate, albeit understandable, that these differences remain largely suppressed in the culture war with the Darwinists. However, I have been quite open about identifying the 'intelligence' of intelligent design with the mind of a version of the Abrahamic God into which the scientist aspires to enter by virtue of having been created *in imago dei*. This claim implies – in a way that has been very controversial in theology but crucial for the rise of modern science – that human and divine intelligence differ in degree not kind. In terms that medieval scholastics of the Franciscan order, notably John Duns Scotus, would have approved, a *univocal* sense of 'intelligent' is attributed to both God and humans, the only difference being that the former possesses infinitely more than the latter. Thus, to say that God 'intelligently designed' reality is to implicate the deity in a process in which humans, however very imperfectly, also engage. Without admitting this semantic point at the outset, the 'intelligence' behind intelligent design would be mysterious and useless to science.

Yet, it is quite a substantial point. It means that we can directly compare ourselves to God and perhaps even chart our progress towards the deity's perfections (Passmore 1970). This conclusion has been strongly resisted by the Roman Catholic Church, which has taken refuge in Thomas Aquinas' counter-claim that all talk about God (e.g. in the Bible) is merely 'analogical', which is to say, to be taken literally only some of the time but figuratively the rest – in proportions to be determined by approved theological authorities. (Protestant detractors coined the word 'equivocal' for this rather flexible approach to the divine word.) Notwithstanding the role that Thomist analogism has played in the development of model building and testing in science, not least in relation to empirical arguments for design in nature (Hesse 1965), it nevertheless presumed that human thought and language are forever limited in their capacity to capture the structure of reality. The

Thomists could not countenance that, in some respects, our minds might literally overlap with that of God's, as in our powers of mathematical reasoning and other forms of *a priori* knowledge. As we saw in Chapter 2, this was something that Aquinas' great nemesis, Scotus, had stressed in his peculiar Franciscan revival of Platonism, which in turn seeded both the Scientific Revolution and the rationalist strain in early modern philosophy. These powers, which Leibniz valorised and Kant ultimately debunked as 'intellectual intuition', were impressive (at least on their face) in allowing us (after Cartesian geometry) to project and (after Newtonian mechanics) to predict vast expanses of physical reality with which we never had – or are ever likely to have – direct sensory contact.

This shift in the status of mathematics from a counting and measuring tool (*à la* Aristotle) to the vanguard discipline of scientific inquiry was part of the modern era's 're-literalisation' of the world, whereby a renewed sense of the continuity of human numerical and verbal expression with the divine *logos* resulted in projects relating to the construction of political constitutions and scientific languages, ideally ones with universal scope and coverage. While at least some ancient Greek schools (e.g. Pythagoreans, Platonists) had presumed the mathematical structure of reality, they did not promote mathematics as a universal medium for the public expression and testing of knowledge claims. On the contrary, they treated maths as a cult activity, one that remained impressive but suspicious to the general public. The same may even be said of writing (e.g. Socrates' distrust of writing in *Phaedrus*). This sociological feature of Platonism has been arguably more powerful than its doctrinal content. However, all of this changed with the mass spread of literacy and numeracy during the Protestant Reformation, which served to dissolve, if not quite annihilate, church authority, which had previously held a monopoly on the technologies of thought through, e.g., dictating and teaching in a Latin that very few understood.

However, inspired by Newton's example (especially as elaborated by his friend John Locke), dissenting clerics did not abandon their ambition to inhabit the divine mind. The 18th century pre-history of experimental psychology, today captured by a catchall 'associationism', included a variety of hybrid scientific clerics such as David Hartley, Joseph Priestley and even John Wesley of Methodism fame (if one counts his interest in medicine), who held surprisingly modern views about the capacity of the mind – understood as the nervous system – to achieve summative, synthetic states of consciousness through what we would now call 'self-organisation'. The resulting heightened awareness could be brought on by scientific work, deep introspection, inspired church singing and teaching, and perhaps even electrical stimulation, not to mention caffeine and alcohol use. It was characteristic of what at the time was first called 'enthusiasm', a coinage that in the original Greek means 'divinely inspired' (Brantley 1984).

Even as associationism became increasingly secular and materialistic in its outlook, it continued to carry its original theological animus, with the human mind portrayed as a kind of chief executive, if not chief architect, of reality, whose capacity to synthesise disparate sensory inputs was modelled on God's giving form to unruly matter in the Biblical account of Creation. Versions of this 'transcendental' view of the mind can be found in Kant, Hegel and their followers, typically with God's existence held in 'suspended disbelief'. While the very idea that divine and human minds might mutually illuminate their workings was already heretical in most Christian quarters, still more controversial was the suggestion that human initiative might be necessary to complete the divine plan. In other words, God and humans are not only mutually illuminating but also mutually dependent.

This view led to a distinctive perspective on the problem of free will, one that Shearmur queries in my avowed sympathy for Priestley and Marx. In the default version of the problem, free will appears to constitute a break in the laws that normally govern nature. The model here is God's miraculous intervention, a capacity that the deity enjoys by virtue of having laid down the laws in the first place. The laws and norms of society presume that humans also have a similar capacity – such that we can be held personally responsible when we are found to have acted against them – though the philosophical basis for this intuition has always been obscure. In contrast, for the associationist, free will and determinism are not opposites but alternative aspects of the same thing. In this casting, humanity's free will enables us to bring (God-like) determination to an otherwise indeterminate situation, be it the multiply firing neurons in our brains or the multiple futures into which the past might be projected (Rivers and Wykes 2008: chap. 3). Without such 'self-determination', we would remain animals driven, as Kant would say, 'heteronomously' by our passions.

In this respect, free will does not interrupt but *produces* determinism: We are entrusted to finish the job that God has started, which amounts to getting our embodied spirits in order so as to get the world in order. (The phrase 'tying up loose ends' is apposite in several senses.) From this perspective, theologians should embrace the prospect of an intelligently designed materialism, as it would allay any doubts that God had failed to discipline matter properly. This certainly helps to explain

Priestley's efforts – amidst of his otherwise deconstructive attitude towards Christian doctrines – to vouchsafe the Resurrection of Jesus, a singular display of God's ability to turn the wayward powers of matter to his decisive advantage. The more that humanity can realise through its own artifice (i.e. technology) the possibilities in nature that God himself had not already done, the more we are engaged in the outworking of Divine Providence, 'building a heaven on earth' (Noble 1997, updated in Gray 2007; for a more positive spin, see Fuller 2010). A 'red thread' to follow this history goes through the value connotations surrounding the concept of *exploitation*, which before Marx was not normally seen as marking a *principled* (i.e. species-based) difference between the moral treatment of humans and non-humans. Even in John Locke, only members of *Homo sapiens* with specific legally protected rights as 'persons' are not subject to exploitation, a view that remained in the framing of the US Constitution.

In this context, classical political economy – especially of the Malthus and Ricardo sort – may be seen as proposing a *rival* philosophy of mind and nature to the relatively passive empiricism championed by, say, Hume. In particular, Malthus and Ricardo, both of whom shared Priestley's Unitarian leanings, characterise humans primarily by their capacities for *producing* rather than *perceiving* the world, whereby nature is treated more as raw material to be worked over than something that we are simply disposed to observe (Young 2000; Cremaschi and Dascal 1996). A good way to appreciate the contrast is in terms of how each secularises the category of *evil* from the original (and near-heretical) science of intelligent design, *theodicy*, according to which the material character of nature seems to require that divine intentions are realised indirectly, rendering 'the best of all possible worlds' compatible with all manner of horrid events. Corresponding to the empiricist notion of *error* is the political economy notion of *waste* – instead of ratios of hits-to-misses, one is concerned with benefits-to-costs (cf. Wise 1989).

Signs of political economy's activist, even constructivist, approach to reality may be found in Priestley's original discovery of oxygen as 'dephlogisticated air'. Even sympathetic commentators continue to focus on his adherence to the outmoded alchemical concept of phlogiston as preventing him from recognising that he had isolated a chemical element (McEvoy 2010). However, by his own lights, Priestley's experiments were prototypes for technologies capable of purifying naturally occurring air to make it better for human consumption. In other words, he saw himself as refining raw matter, not revealing a pre-existent reality. Indeed, he located the benefits of dephlogisticated air in the marketing of soda water as an elixir rather as an opportunity to reconstitute the foundations of chemistry (Johnson 2008: chap. 4).

Together these historical observations provide the rudiments for a unified science of intelligent design that divides into two main branches: divine artifice (aka biology) and human artifice (aka technology) – the former literally considered as a superior version of the latter, or the latter an inferior version of the former, or perhaps the two artifices co-produced in some way, all depending on one's theological starting point. But the literalness of the comparison, whichever way it goes, is fundamental. It implies *inter alia* at least some commitment to Platonism, whereby one might say that God and humans might instantiate the same form in matter to varying degrees of realisation. Hints of the foundations of such a science may be found in the regulative ideals of reason found in Kant, which William Whewell and successive philosophers in the 19th and 20th centuries expanded into second-order normative principles of scientific inquiry, such as intelligibility, unity, simplicity, fecundity, breadth, depth, etc. Software engineers – who appear to be the main constituency for intelligent design theory without an obvious theological position these days – speak of 'system quality attributes' as latter-day versions of these notions (Russel 2009). Such principles are of an abstract psychological nature, defining the parameters of our cognitive horizons, which at the limit result in the ultimate object of knowledge, the point where human and divine minds completely coincide. The foundations of knowledge would then be a kind of divine psychology, which is not so very far from James Ferrier's 1854 English coinage of 'epistemology' (Passmore 1966: 52–3), important aspects of which carried through to artificial intelligence research and general systems theory in the second half of the 20th century (e.g. Simon 1977; Rosen 1999), though to a lesser extent the branch of academic philosophy today called 'epistemology'. In short, human minds would provide a fallible but corrigible basis for modelling God's mind.

I wrote *Dissent over Descent* (Fuller 2008a) with an eye to reconnecting theology and science, which is to say, with an emphasis on the *intelligence* behind 'intelligent design'. However, most of the debate in this field has so far centred on the relevant sense of 'design', not least because of the seemingly ineliminable character of design-talk in contemporary biology, which if anything has grown stronger as the work of the discipline has migrated from the field site to the molecular laboratory and computer simulator. In the book, I question the curious Neo-Darwinian assumption that because certain evolutionarily stable changes can be made to an organism or a population under the highly constrained (aka intelligently designed) conditions of an experiment or a simulation, it follows that those changes must have also routinely occurred in a more spontaneous fashion for all forms of life over millions or billions of years. Such an assumption would never go unchallenged in the social sciences for our own species over a few hundred or thousands of years (Fuller 2008a: chap. 5). Here we see the confidence – some might say hubris – that the combinatorial possibilities of complex molecules conjoined to an indefinitely old Earth have bred in the Darwinised mind.

This point recalls the rhetorical blunder that intelligent design champion, the biochemist Michael Behe (1996), committed – which came back to haunt him as an expert witness in *Kitzmiller* – when he claimed that bacteria are so 'irreducibly complex' (i.e. they were designed fit-for-purpose) that they *could not possibly* have acquired their mobility function through random mutation and natural selection. Of course, Behe's charge can be easily rebuffed if Darwinian evolution is a probabilistic process that is allowed enough time to run its course with some minimal path dependency built into the outcomes. In that case, there are many possible scenarios by which bacteria could have acquired their form. But instead of questioning whether Neo-Darwinism could provide *any* account of bacterial mobility, Behe should have asked how they would determine *which* account was true, thereby calling the theory's *falsifiability* into doubt. In a sense, then, anti-Darwinist appeals to statistical arguments are bound to prove inconclusive – that is, unless it is also shown (say, by radiometric interpretations of fossils) that the Earth is significantly younger than it is normally thought to be.

Other than whether to take biology's pervasive design talk literally, the most controversial question relating to design in nature concerns the 'units of design': Exactly what sort of thing is supposed to be, in the intelligent design jargon, so 'irreducibly' (Behe) or 'specifically' (Dembski) complex as to imply a designer? William Paley, the historic standard-bearer for intelligent design theory – largely because of the negative example he provided for Darwin – proves to have been a transitional figure in the history of design thinking. To be sure, Paley retained the ancient Aristotelian *typological* perspective, which presumes that every normal member of a recognised species is designed (or 'pre-adapted') for its environment. However, Paley supplemented this with a *populational* perspective, indebted to his fellow cleric Malthus, which justified differential rates of survival – especially amongst various nations and classes of *Homo sapiens* – as providing at least indirect lessons in the conduct of life. Thus, Chapter 26 of Paley's *Natural Theology*, entitled 'On the

Goodness of the Deity', is devoted to a defence of Malthus' controversial (at least amongst Christians) for his call to end Poor Laws as a futile exercise in resistance to divine will. Darwin not only abandoned the typological in favour of the populational side of Paley's scheme, but he also divested the populational side of its link to theodicy, reflecting Darwin's unwillingness to credit a Creator who would allow so much wasted life. For Darwin, ever the Epicurean, suffering *as such* is evil, even were it to come from a deity whose ultimate sense of benevolence is brought about by such cruel means as mass extinction.

Darwin's reasoning here anticipates somewhat Popper's arguments against millenarian utopianism, which would sacrifice many lives in the name of some perfect vision of humanity (Popper 1946). Any powerful being willing to use such means would not be worthy of one's allegiance, however noble the aim. However, Jesuits attempted to mitigate such reasoning in the Counter-Reformation by arguing that the material character of Creation generates so much suffering merely as an *unintended consequence* of divine agency. Of course, this argument has enjoyed its own chequered history as secular theodicy mutated into classical political economy, of which Milbank (1990) has provided a 'radical orthodox' critique.

It is worth observing that while historians and philosophers of biology such as Ernst Mayr (1970, 1982) and David Hull (1989) have pointed to Darwin's definitive shift from typological to populational thinking about species as constituting the biggest conceptual breakthrough in the history of biology, the shift was inspired by the quantitative precincts of social science – specifically what was first called in the 17th century 'political arithmetic', which by the 19th century was called 'social statistics' (Porter 1986). This is a crucial but underexplored part of the backstory to contemporary intelligent design theory. The heavy involvement of theologians and theologically-oriented thinkers in this tradition – not least the Reverend Thomas Bayes – should remind us of the substantive role that univocal comparisons between divine and human qualities have played in the rise of modern science. In this respect, the history of statistics can be told as one long attempt on the part of humans to second-guess divine governance for purposes of drawing lessons for human governance. The central source of disagreement in this tradition has been whether long-term statistical tendencies should be taken as themselves the outworking of Divine Providence or signs of an open situation that requires human intervention for God's work to be done. Responses to Darwin's own work parallel the two traditions from a humanist, if not quite atheist, perspective. The former captures the *laissez faire* mentality promoted by Herbert Spencer, while the latter corresponds to Francis Galton's eugenics campaign, which reintroduced many of the normative judgements about individuals that had been associated with typological thinking (e.g. variation in traits must be properly distributed). As social policies, the former suggested tolerance if not neglect, whereas the latter implied legislation if not coercion.

The larger lesson of the history of statistics for intelligent design is that claims about design are necessarily holistic. Nothing is well or poorly designed as such but only in relation to some overarching ends. Even if, as Paley was inclined to argue, a particular organism appears especially well designed for its native habitat, the organism's design features must be seen as part of an overall plan in which they can be shown to be functional. The proper analogue to his famous found watch on the heath is not an organ or organism but the universe as a whole. In other words, there is an implicit ecological dimension to intelligent design theory, one that Priestley, for example, exploited when he provided one of the first accounts of photosynthesis – though in characteristically 18th century form, Priestley's account of the process was mechanical, akin to a factory's production line (Johnson 2008: chap. 2).

To be sure, this way of seeing things runs counters to today's intuitions, which (*pace* Paley) does *not* regard 'design without a designer' as an oxymoron. Thus, we routinely explain the heart's design in terms of its function in our bodies without having to explain how our bodies came to be as they are. Interestingly, the heart was probably the 17th century's favoured example of 'irreducible complexity' in empirical proofs for an intelligent designer. But those who appealed to the example, notably the great Cartesian Malebranche, embedded their arguments in an Abrahamic natural theology, with its prior expectation that *human* bodies would be especially well designed because of our supreme status in nature (Pyle 2003: 163–4). There were no such expectations about the organs of lower creatures, which obviously survived but in a lesser state of being, which is to say, not so well designed – a sensibility that lingers in the folk idea that most animals are 'dumb'. A subtle achievement of Darwinism was to overturn these expectations, so that now we presume that all reproductively active species are equally well adapted to their environments, and so their organs can be presumed equally well designed. Such is the blind justice delivered by natural selection. In other words, by abandoning the idea that the human body is the benchmark of God's handiwork, it is no longer necessary – or perhaps even possible – to infer the designer's signature in order to account for the presence of design in the bewildering array of species that have inhabited the planet.

In contrast, to someone living a couple of generations before Darwin, such as Priestley or Paley, the ultimate normative status of certain states of affairs – and hence what renders them justified – remains an open question until the completion of the divine plan, the 'Final Judgement', as Christians like to say. Thus, apparent imperfection in the design of some organism or other aspect of nature may be indicative of a mere lack of self-sufficiency, a work in progress – globally speaking, a means whose divinely appointed end has yet to be discovered. Although the watch that Paley imagined us to have found on the heath was in working order, it need not have been for us to infer that it was intelligently designed. Had we found instead a broken watch, we could have proceeded in one of two ways: either seek the other pieces to restore its original integrity, or conclude that something inhibited the watch's completion. Both prospects suggest that the watch met a violent fate, the latter potentially involving the entire society for which it was designed. However, in both cases, the watch's discoverer – as herself an intelligent designer – is well-positioned to repair the watch.

This fundamental uncertainty about underlying design principles – whether they are to be restored or fully realised, as in the case of Paley's watch – has been instrumental in generating the host of conservative, liberal and radical attitudes that have marked humanity's relationship to the rest of nature in the modern period. Each has offered rather different takes on how the future follows from the past. And just as science has tracked theological positions – both critically and dogmatically – it has tracked these secular ideological ones as well. For better or worse, and perhaps surprising to all concerned, social engineering is a secular offspring of intelligent design theory. In the remainder of this chapter, I shall consider two heretical scientist-theologians who leveraged a version of intelligent design theory into a radical secular worldview that would see humanity come to realise its consummative role in the unfolding of the divine plan – two versions of *Theology 2.0*. The first, Joseph Priestley, has been already highlighted in these pages. The second, Pierre Teilhard de Chardin, was a Jesuit palaeontologist, part of the original 1923 expedition that discovered the Peking Man, a 500,000 year old ancestor of *Homo Sapiens*, and later an inspiration for the evolutionary humanism of Julian Huxley (who was responsible for getting Teilhard's papally proscribed works into print after his death) and Theodosius Dobzhansky (1967).

4 Joseph Priestley's Theology 2.0: The completion of Newton's Unitarian project

Joseph Priestley (1733–1804), now known mainly as the unwitting discoverer of oxygen, was probably the most important if not most influential English-speaking intellectual of the 18th century. He was certainly the most interesting. (In making these claims, I mean to include Berkeley and Hume, but not Locke, as 18th century figures.) However, like so many intellectuals, Priestley is remembered, if at all, in fragments – more often for the extremism, oddity, or sheer error of his views than their actual content, import, or impact. Nevertheless, over the years there has been a small but steady stream of scholarship relating to Priestley's career, typically in search of an elusive underlying unity. John McEvoy (2010) and Robert Schofield (1997, 2004) have been the most distinguished contemporary contributors to this task.

Despite his prodigious output, no single work of Priestley's captures the full measure of his thought. His more scientific works are either advocacy histories or idiosyncratic phenomenologies, while his more philosophical and theological tracts were typically dashed off in response to controversies of the day. These literary features have seriously impeded Priestley's reception in our own time, when works are expected to slot into pre-ordained disciplinary categories and, in particular, 'philosophical' works are expected to be pitched at a level that errs on the side of making one's reasoning too explicit (even if boring to the original readers) and one's target too implicit (even if confusing to the original readers). In this way, a 'philosophical' argument can be examined in suspended animation. Priestley defied posterity – so far to his disadvantage – by communicating to his target audiences as directly as possible. But there have been attempts to redress the balance in recent years (Johnson 2008; Rivers and Wykes 2008).

Steven Johnson's *The Invention of Air* (2008) is a popular history that succeeded to bring Priestley to the attention of readers of *The New York Times Book Review* (2 January 2009). Johnson consulted archives on both sides of the Atlantic and routinely grounds his narrative in bits of correspondence that provide glimpses into why so many intellectuals, politicians, and journalists took Priestley so seriously even as they denounced him for unconscionable radicalism, such as when Priestley supported the French Revolution partly to inspire a similar republican outbreak in Britain. Although Priestley, no less than Newton, embedded his scientific investigations in a Biblically inspired metaphysical framework, the character of Priestley's theology is not as attractive today as it was, say, 150 years ago. Whereas Newton continues to intrigue readers with his behind-the-scenes attempts to divine the nature of physical reality from the pages of Scripture, Priestley disturbs the average churchgoer with his relentless attack on ecclesiastical authorities of all denominations. Yet, the theological position of the two figures was largely the same, Unitarianism, which effectively carried the Protestant Reformation to its logical conclusion by removing the last vestige of mediation between God and humanity, to wit, the doctrine of the Holy Trinity. The difference between Newton and Priestley was largely a matter of style that reflected the difference that a hundred years made: What Newton had to do in secret, Priestley did very much in the open – though admittedly it resulted in the torching of his Birmingham home and consequent exile to rural central Pennsylvania.

Johnson's narrative is centred on Priestley's long-standing friendship with the US founding fathers, especially Benjamin Franklin and Thomas Jefferson, whose intellectual profiles resembled Priestley's own, even down to the fondness for alcohol and caffeine – the recurrence of which in this story would be worthy of treatment by a 'neurohistorian'. Priestley's pioneering *History and Present State of Electricity* (1767) was responsible for manufacturing the image of Franklin's kite flying as the moment when humanity finally harnessed electricity from the heavens. Later Priestley would prod Jefferson to produce a version of the New Testament that stripped the life of Jesus of all superstitious elements – resulting in a devotional text the size of a long pamphlet. In return, Jefferson helped to provide safe passage for Priestley to the United States, where he turned down the opportunity to be the founding chair of the University of Pennsylvania's chemistry department in favour of railing against the newly passed Alien and Sedition Acts, which threatened to undermine the civil liberties on which the new nation was based (Rivers and Wykes 2008: chap. 7).

While dutifully recounting Priestley's rivalrous correspondence with Lavoisier over the explanation of 'dephlogisticated air', Johnson casts a fresh eye on what was at stake for Priestley. Rather than seeking to identify the elements that might establish the foundations of a new science of chemistry (which is what Lavoisier thought he was doing), Priestley understood his experimental isolation of oxygen as a refinement of the earth's raw materials, a technological innovation consonant with what we would nowadays call 'natural capitalism', whereby the most efficient mode of production is also the one that leaves the world in ecological balance. Here Johnson deftly connects Priestley's interest in the elixir-like qualities of oxygen – marketed in the 19th century as soda water – with his discovery of the carbon cycle in photosynthesis. From a public relations standpoint, Priestley had re-branded plant life from passive ingestors of soil nutrients to active manufacturers of a key environmental condition that enables all life to flourish. Given the scientific honours bestowed in his lifetime (despite his radical politics), Priestley's revisionist approach to plants should perhaps be seen as comparable to evolutionary psychologists who today claim to have demonstrated that animals display forms of cognition above mere sentience.

Perhaps surprisingly for a popular work, *The Invention of Air* provides clues for piecing together Priestley's scientific worldview, which may be encapsulated as follows: A providentialist natural theology supplies the explanatory framework within which what would be normally called scientific discoveries are understood as prototypes of technologies through which our own godlike creative powers enable us to perfect the divine plan – which is to bring about 'a heaven on earth'. The view is not unfairly called 'Christian Materialism', and resonates strongly with the Mormon idea that humans are potentially 'latter-day saints'. (Both Priestley and Mormons share a literalist interpretation in the Resurrection of Jesus, which they then seek to understand scientifically.) Put this way, it is not surprising that Friedrich Engels, in *Ludwig Feuerbach and the End of Classical German Philosophy* (1888), regarded Priestley as a truer precursor to his own brand of historical materialism than French *philosophes* like Baron D'Holbach who appealed to our material nature as Epicurus might – that is, to negate the prospect of self-transcendence. Paradoxically perhaps, Priestley appeared to capture the 'spirit' of the Marxist project much better than the atheistic materialists of his day.

The essays collected in *Joseph Priestley* (Rivers and Wykes 2008), though largely by and for specialists, would do the most good in the hands of those seeking general guidance in how it has been possible to seek God through science and thereby provide meaning to the human condition. While this broader concern seeps through most of the essays, they are written in the spirit of scholarly apologetics for devoting so much attention to such an outlier from the canonical narrative of the history of modern philosophy, science, and culture more generally. Priestley's outlier status is ultimately grounded in his refusal to draw sharp boundaries between various bodies of knowledge that in the modern era have become increasingly distinct – say, philosophy, science, politics, and theology. This point comes out most clearly in Chapter 6, which focusses on Priestley's *Lectures on History and General Policy* (1788), which drew on courses he gave at Warrington Academy, a school for religious dissenters near Manchester with which Priestley was associated from 1765 to his emigration to the United States in 1794. Its pupils included Thomas Malthus and Christian Carl André, a naturalist who went on to found the Moravian Academy of Sciences in the Hapsburg Manchester, Brno, later home to one Gregor Mendel, who leveraged Priestley's 'natural capitalist' views of plants into the first modern scientific theory of heredity (Wood and Orel 2005; cf. Fuller 2008a: chap. 2).

Priestley's pedagogical innovations were so fundamental that their radical character in late 18th century Britain can be easily overlooked today. He had students learn about the nature of language by studying English grammar, rather than Latin or Greek. He put modern history on the curriculum when only classical history had been taught. Moreover, Priestley urged a forensic approach to historiography – to wit, that the original documents and artefacts ('material evidence') be consulted whenever possible. Priestley's clear target here was Edward Gibbon, whose magisterial *Decline and Fall of the Roman Empire* (1768) was, in effect, a very learned allegory constructed from a critical reading of classical sources for purposes of offering moral instruction to the present, given what Gibbon presumed to be recurrent tendencies in human nature. Such 'philosophical history' was standard fare at the time, with Gibbon distinguished simply by his narrative's anti-Christian undertow. For his part, Priestley, shared neither Gibbon's cyclical view of history nor his tendency to use the past as a pretext for talking about the present. After all, if history is heading in some providential direction, as Priestley thought, then the past should not cast such a heavy shadow on our understanding of the present. Indeed, the past may tell us more about the errors we have made (and hence should avoid) than where we are heading.

In today's terms, Priestley might be seen as akin to those in cultural studies who are less concerned with the scholarly authority of historical sources than their proximity to the events that one seeks to understand. In this context, journalistic accounts and popular representations become epistemically luminous. Other aspects of Priestley's pedagogy pointed in this direction. He famously introduced time-lines into the teaching of history but less well known is that he selected and arranged historical personages on the basis of fame rather than merit, since (according to Priestley) judgements of the former tend to be more informative than those of the latter. Equally remarkable for his time, when experimentation still appeared to be a somewhat esoteric way of obtain knowledge, Priestley encouraged children to play with scientific apparatus and then reflect on how their discoveries reflected the overall design of nature. He was more interested in children learning to see nature's order in the artifice of the laboratory than the actual correctness of their conclusions. Little surprise, then, that Priestley disapproved of Hume's sceptical *Dialogues concerning Natural Religion,* given their ambiguously negative stance on the scientific detectability of design in nature.

Priestley scholars are divided over whether there is a unifying principle to his thought (Rivers and Wykes 2008: chaps. 2–3). It is clear that like the other great polymaths of history – and Priestley is neatly placed in terms of interests, disposition and chronology somewhere between Leibniz and Goethe – he spread himself too thin. Nevertheless, Priestley's centrifugal tendencies were held together by a commitment to the specific brand of associationist psychology that he had inherited from the physician-turned-dissenting-minister, David Hartley (1705–1757), whose *Observations on Man* (1749) Priestley abridged and edited for late 18th century publication. Hartley radicalised the empiricism of John Locke, himself a trained physician of non-conformist religious leanings. Indeed, he offered the first systematic neuroscientific account of mental life by accounting for the centrality of ideas in Locke's epistemology in terms of the summation of nerve vibrations distributed throughout the body. Most significantly, Priestley adopted from Hartley the principle that mental association is governed by the contiguity, not the similarity, of nerve impulses. Accordingly, ideas are synthetic products of fused impulses, an image that the German philosophical tradition would later use to model dialectical processes both within and between humans. It is clear that Priestley himself already envisaged associations at multiple levels. His Warrington curriculum fostered associations between reading and pleasure, while he promoted the free transit of ideas among mature individuals as eventuating in an improved collective intelligence – what Hegel and his Marxist followers subsequently identified as the conversion of quantitative to qualitative change. Priestley even anticipated Marx's view of the politically revolutionary potential of this transition.

With the benefit of hindsight, Priestley's optimism appeared to presume that a sufficient number of properly educated people could reverse the statistical tendency of societies 'regressing to the mean' by exemplifying how one resolves disparate considerations for oneself. The preference for contiguity over similarity as the principle of association presupposed that if people are not specifically encouraged to interrelate their own personal experiences for purposes of public expression, they might simply copy the self-expression of dominant members of their society. From an associationist standpoint, it would mean that other people's experience had colonised their own minds – perhaps to such an extent that one's own lived experiences are systematically discounted in public self-representations. Priestley quite reasonably wondered how such a regime of self-censorship could promote the search for truth. He wanted individuals to learn how to make trade-offs for themselves without expecting that, under similar circumstances, others would make the same trade-offs. This talk of trade-offs serves as a reminder that for Priestley an educationally enhanced commercial ethic provided a better guide to citizenship than simple obedience to established political authorities.

Let me conclude by observing that Priestley enjoyed enormous respect on both sides of the Atlantic in the first 150 years following his death. He was routinely regarded as not only the discoverer of the oxygen but also an icon for the wide-ranging citizen-scientist, which suited the self-image of chemists throughout the 19th and early 20th century. However, all of this changed with fellow chemist James Bryant Conant's famed case histories of the experimental sciences, which formed the basis of the postwar Harvard General Education in Natural Science course in which one Thomas Kuhn served as teaching assistant. Conant and Kuhn (in *The Structure of Scientific Revolutions*) placed a canonical stamp on Lavoisier's victory over Priestley for the discovery of oxygen, something that until then had been subject to much contestation (Fuller 2000b: chaps. 3–4). It is difficult to resist the view that Priestley's explicit marginalisation was intimately connected to his refusal to disentangle his 'science' from his radical politics and theology. In that Cold War context, Priestley had the audacity to suggest – in both word and deed – that the pursuit of science could constitute an all-encompassing lifestyle that as a matter of principle shuns service to established authorities.

5 Teilhard de Chardin's Theology 2.0: The completion of divine creation itself

Pierre Teilhard de Chardin (1881–1955), the maverick Jesuit palaeontologist, is also a futurist out of fashion – indeed, on first appearances, a typical casualty of futures studies: the grand visionary of a future past, a man whose sense of tomorrow was too well grounded in the knowledge and hopes of his day. Already in 1965, Stephen Toulmin had judged Teilhard's *magnum opus*, *The Phenomenon of Man* (Teilhard 1961) 'wish-fulfilment' (Toulmin 1982: 113–26). Nevertheless, Teilhard figured prominently in the first decade of the journal *Futures*: 26 of the 42 citations to his work can be found between 1969 and 1979. The 1980s and the 1990s each produced five citations, and so far this decade has produced six. A point of comparison is Lewis Mumford's citation history in this journal: 52 overall and distributed relatively equally across the decades, with only 15 in the first decade.

It appears that Teilhard projected a future that seemed feasible in its time but happened not to be the one that was realised. It is now hard to believe that Teilhard, whose views were deemed so dangerous that he was banned by the Pope from taking up a chair in the Collège de France, had been such a popular figure in the early days of *Futures* (Brett-Crowther 1981). But *Futures* was hardly alone. 1969 witnessed the depiction of Teilhard's plight in the Hollywood blockbuster, *The Shoes of the Fisherman*, in which Oskar Werner upstaged Anthony Quinn in a supporting role as the heretical but sympathetic Father David Telemond. In the following decade, I recall vividly Teilhard's ideas being discussed in hushed tones by my Jesuit prep school teachers in the context of anti-war protests, Marxist existentialism and liberation theology, all of which – as it turns out – were in their death throes.

Teilhard's decline can be easily explained: His most distinctive theses, which made him appear so radical in the 1960s, have been rejected as beside the point, if not altogether wrongheaded. In the rest of this section, I perform an autopsy on the demise of Teilhard's reputation and look for possible signs of recovery. I consider two matters in sequence as opening moves towards a rehabilitation of Teilhard's general orientation to the future. First, I identify the political and scientific currents that originally rendered Teilhard's vision obsolete. Here I observe that the relevant currents appear to have reversed in recent years, offering new hope for a Teilhardian revival. Second, I show how principal-agent theory in political science can be used to flesh out the details of Teilhard's creative evolution as a paradigm for futures studies. In this context, God is the 'principal' and humanity the 'agent' whose collective effort is needed to realise the divine plan. It is left for the reader to decide whether the progressive scientific vision so entailed requires the accompanying theological scaffolding.

Teilhard's precipitous decline into obscurity is traceable to three awkward features of his worldview that perhaps appear less awkward today:

(1) Teilhard believed that science was humanity's species-unique and divinely privileged vehicle for knowing God, which amounts to becoming – or fully realising – God. To be sure, other life forms

participate in divine creation through their sensory ties to the material world, but only humans are capable of transcending that relationship to provide order and ultimate meaning to the whole of material reality. (In this respect, Teilhard provides a sharp rebuke to religiously inspired naturalism of the sort associated in the early mid-20th century with George Santayana.) Teilhard's position is nowadays dismissed on both scientific and religious grounds, which Stephen Jay Gould (1999) summarised in terms of science and religion constituting 'non-overlapping magisteria'. On the religious side, Teilhard seemed only to grasp the human side of 'transcendence', which involves going beyond one's natural limits, but not the divine side, which involves surrendering oneself to another, the stuff of *faith*, the point at which religion classically takes over from science. On the scientific side, biologists nowadays generally uphold an indefinitely divergent view of evolution, in which *Homo sapiens* may simply be one branch, doomed to termination once selection pressures get the better of us. It is perhaps no coincidence, then, that the evolutionists who promoted Teilhard's case most strongly – Theodosius Dobzhansky and Julian Huxley – were also eugenicists, albeit of a cautiously liberal disposition. If nothing else, eugenicists thought they could steer the course of evolution (Fuller 2007b: 136–9). However the tide may be turning in Teilhard's favour against Darwin's indefinite sense of species differentiation. In particular the recent revival of developmental biology (i.e. the 'evo-devo' approach, cf. Carroll 2005) and more explicitly theistic forms of convergent evolution (Morris 2003) stress the recurrence of structures in later organisms that are functionally equivalent to those found in earlier, often extinct ones. In effect, they are rediscovering a sense of focus, if not purpose, in nature that Darwin never could have imagined.

(2) Teilhard believed that the Earth was evolving towards becoming a single 'hominised substance', almost the exact opposite of, say, James Lovelock's Gaia Hypothesis, which has animated much recent ecological thought. It is now increasingly common to treat *Homo sapiens* as, at best, *primus inter pares* amongst species and, at worst, a blight on the biosphere. Gaia presupposes Darwinian ethical scruples, whereby no species should require so much for its survival that it crowds out other species. This line of thought could not be further from Teilhard's rather absolute views about humanity as the crown of creation. Here he recalls Lamarck and that most ambitious of Enlightenment *philosophes*, the Marquis de Condorcet, who similarly

looked forward to when humanity would make good on its Biblical entitlement to earthly dominance, which he interpreted quite literally in terms of our becoming the planet's sole inhabitants (Fuller 2006b: 162-4). From that standpoint, all the other creatures are transitional forms or prototypes for modes of being that we might come to understand, improve and incorporate into our life-world. To be sure, much of this 'incorporation' occurs by the normal biological means (e.g. food and shelter) but increasingly it is expressed as technology, perhaps even what Richard Dawkins has called the 'extended phenotype': We adopt and adapt design features of the non-human world. In this respect, Teilhard was a fellow-traveller of the disciplines that originally inhabited the interface between biology and technology: bionics and biomimetics, which inspired humans to surpass the birds as creatures of flight (Rosen 1999: chap. 19). Of course, with advances in biotechnology, direct incorporation of the non-human into the human through xeno-transplantation is becoming more common. Transhumanists anticipate our gradual evolution into cyborgs, a prospect that in principle would have been acceptable to Teilhard, assuming that the enhanced capacities of our cyborg successors are ones that humans already possess, albeit to a lesser extent.

(3) Some futurists have tried to identify Teilhard's ultimate realm of being, the 'noosphere', with the internet, understood as an emerging global electronic brain, the first genuinely universal medium of intellectual exchange that will provide the basis for a step-change in the human condition (Anderson 2003). While Teilhard may have been inspired by H.G. Wells' telecommunications superorganism, the 'world brain' (Rayward 1999), little of the actual internet would have appealed to him. In particular, the ersatz noosphere known as the 'blogosphere' would strike Teilhard as a nightmare version of Leibniz's monadology, according to which reality is constructed through a pre-established harmony (i.e. weblinks) amongst monads (i.e. blogs), each of which reflects the totality of reality from its own autonomous perspective. Rather, Teilhard would have been most intrigued by what many consider the internet's most disturbing feature, namely, the propensity of its more frequent users to transfer affection and even self-identity from their physical bodies to virtual realisations, or 'avatars', which enable various sorts of interactions in a common medium that would otherwise not be possible. This transference on a mass scale, combined with a mechanism for the avatars' self-reproduction, if not perpetuation, could

> count as the quantum leap in evolution that Teilhard projected from the biosphere to the noosphere. It is a prospect that science fiction has already explored in depth (e.g. Stephenson 1992). But a convincing backstory is needed to alleviate the anxieties surrounding such 'disembodiment'. The following is a sketch: Once modes of legitimate succession started to be forged along artificial rather than natural lines with the advent of the corporation (i.e. the 'artificial person', or *universitas*) as a category in Roman law, the path to the noosphere started to be paved. In this respect, Thomas Hobbes deserves credit for having generalised this eccentric legal category into the signature human achievement with his conception of the 'social contract'. This too led to a transfer of affections and self-identification over time, as claims of patriotism trumped those of paternity. Computer avatars may simply be one more stage in that process (Fuller 2006b: 201–5).

Any rehabilitation of Teilhard must begin by taking seriously that he was a 'creationist' in a sense that places him squarely in the monotheistic tradition. This tradition, which flows from Genesis, postulates a God who needs to create in order to be fully realised. Matter, presented as inert or chaotic is, in any case, radically different from God yet also the medium through which the divine project is completed. As periodically observed in these pages, a special branch of theology, 'theodicy' (literally, 'divine justice'), was established to tackle this problem in a concerted manner, from which modern ideas of optimisation in engineering and economics have descended (Fuller 2008a: 171–81). Theodicy gradually went out of fashion – or, rather, migrated to these more secular sciences – once Kant demonstrated the futility of trying to fathom God's *modus operandi*. I shall attempt to resurrect theodicy one more time in the closing chapter of this book.

Gottfried Wilhelm von Leibniz, who wrote the most famous book on theodicy, cast the discipline as addressing the nature of creation as the ultimate metaphysical problem: 'Why is there something rather than nothing?' Leibniz's solution lay in an appeal to the principle he called 'the identity of indiscernibles'. In other words, two things are the same if there is no way to tell them apart (in space). From this principle, we might then argue that there is an ordered universe (the 'something') in order to indicate the difference that God's presence makes. Were there no such universe, there would be no reason to think that God exists. Whatever one makes of Leibniz's argument as either a proof of God's existence or a justification of an ordered universe, it highlights God's dependency on creation as the grounds for his authority. Yet, matter's otherness renders creation unruly by nature. It has, so to speak, a mind of its own. In that respect, God is forced into a 'principal-agent' relationship with creation. Thus, Teilhard's distinctive brand of 'creative evolution' can be understood as a theory of the unfolding of that relationship.

God delegates power (via the creative word, or *logos*) to various organisms, humans most notably, to execute the cosmic plan in a resistant material medium. This delegation establishes God as the principal and his creatures as agents. As a good Lamarckian, Teilhard believed that species lineages are rather like lines of thought routinely tossed off by a very creative thinker, which in this case is God rather than some exceptional human genius. Most never reach full realisation but they are all 'groping' (*tâtonnement*) towards 'the omega point' (Fuller 2008a: 102–7). The secularisation of Christianity from the Protestant Reformation onward was driven by the increasing literalness with which philosophers and theologians took the 'devolution of the divine corporation', which essentially turned the agents into themselves principals, who in turn delegated to still other agents (Schneewind 1984). It justified a transfer of sovereignty from the Pope, as the representative of God on Earth, to individual churches, states and ultimately biological individuals who realise the divine plan through their spontaneously self-organising and mutually recognising exercise of intelligence. This process culminated in a host of doctrines ranging from deism through humanism to outright atheism, including the so-called invisible hand of capitalism, each motivated by the prospect that the divine principal had ceded *all* his authority to human agents.

Principal-agent theory was invented to cope with the increasing distinction between the interests of owners and managers of business firms. Discussions of this theory have centred on problems relating to how owners can provide the right incentives for managers to do their bidding, given that the two groups tend to be motivated and informed rather differently. Of course, these problems also apply to the theological case, where God is the principal and humans the agents. But here matters are rarely discussed so bluntly. Theologians tend to focus on how the human agents might continue to possess free will while executing the divine principal's plan, rather than how such agents might be motivated to execute the principal's plan successfully. Concerns for human dignity thus outweigh those of divine efficacy.

John Stuart Mill sarcastically captured this situation as implying the existence of a 'limited liability God', whereby in exchange for free will, the human management team cannot blame the divine owner for failing to execute an overall plan for which the team has been specifically employed. At times like this one wishes that theologians took literally the Genesis claim that humans are created in the image and likeness of God. One implication would be that the more we learn of the history of humanity, including how the human mind works, the more we shall gain insight into divine intelligence, whose intentions we might then come to realise more fully, given God's declaration that he is the most perfect version of ourselves. If there is a Teilhardian imperative for 21st century futures research, this is it. In what follows, I consider a futures-oriented example of what creative evolution looks like as a version of principal-agent theory.

Consider a future in which people can control – by whatever natural or artificial means – their exact physical longevity. How would our successors explain their having reached that point? They would probably appeal to previous ideas and empirical clues advanced in the name of achieving such a goal. However, the allure of the goal would have been often conveyed through academically peripheral channels like science fiction novels that permit authors the divine luxury of designing imaginary worlds that, at the very least, keep the target in vivid view when more 'realistic' minds attend to other matters. Indeed, as we saw in Chapter 1, this was the spirit in which H.G. Wells had campaigned (unsuccessfully) for the founding chair in sociology at the London School of Economics (Lepenies 1988: 145–54). These realistic minds – scientists working on conventional biomedical solutions to specific human ailments – would have become acquainted with the more fantastic texts, perhaps through avocational reading or film adaptations. In any case, the content of those texts would have lodged in the backs of their minds, informing their private justifications of their own limited contributions, unless of course those contributions appear to have conformed to what they have encountered in some scientifically sanctioned literature.

This story suggests the image of a virus that lies dormant until the conditions are right to activate it, or to enable it to switch on the appropriate gene. The precedent set by the earlier strictly non-scientific 'viral' works would include explaining the significance – if not some of the detail – of what turns out to have been the subsequent scientific achievement. After all, how do you tell the difference between a striking but basically one-off event and a proper 'discovery' on the basis of which long-term lessons may be drawn? The answer is to connect the event in question with earlier aspirations that reveal it to be the solution to a long-standing problem, the implications of which have been already largely anticipated by an earlier literature.

Unfortunately, at least for authors who aspire to inclusion in the history of epistemology, fiction (including science fiction) is rarely given its due. Here 'philosophy', understood as a genre of writing, occupies a respectable middle ground for scientists who do not wish to admit that their discoveries are best explained by literary forms redolent of whole cloth fabrication. Nevertheless, it is curious just how willingly scientists continue to trade on the fragmentary and ambiguous speculations of a pre-Socratic philosopher like Thales to justify the search for the ultimate nature of matter. Were this project to be successful, it would have happened by means that old Thales could never have imagined. At the same time, more explicitly literary precedents for that discovery along the way would be deemed too 'fictional' to be credited, precisely because of the explicit way they preclude empirical accountability.

In sum, the creative evolution of a scientific discovery is clear: An essentially unpredictable event is inserted into a narrative that is 'overdetermined' – that is, one informed by prior expectations so as to be, in principle, realisable in many ways. The result is a generally recognised breakthrough in knowledge. Thus, discovery favours the prepared *collective* mind. Such overdetermined accounts presuppose a collective historical memory on which to draw that stresses *unfulfilled* aspirations over successive generations, an instance of Teilhard's 'groping' (*tâtonnement*). In this connection, humanity is distinctive in its studied cultivation of endless yearning and hence regular disappointment, a dynamic condition that exists comfortably within Darwin's struggle for survival but corresponds more precisely to Hegel's struggle for recognition (Fuller 2006b: 107–17). An interesting lead for continuing this line of thought would be through what the Gestalt psychologists originally identified as the Zeigarnik Effect, only now raised to a higher-order level (Zeigarnik 1967). Thus, whereas Zeigarnik focused on our persistent memory of unfinished tasks that unconsciously hastens their completion, Teilhard treated divine creation as the ultimate of such tasks.

5

Conclusion: In Search of Humanity 2.0's Moral Horizon – Or, How to Suffer Smart in the 21st Century

The argument of this book's concluding chapter takes literally the proposition that whatever else an 'enhanced humanity' might be, it somehow brings us closer to a divine standpoint – in the specific sense that our properties become more like those of the Abrahamic deity. Here I continue to draw implications from John Duns Scotus' doctrine of 'univocal predication', first introduced in Chapter 2. The chapter is anchored in the traditional theological barrier to our comprehension of God's ways, namely, the pervasiveness of suffering, often personified as 'Evil'. I note that modern redistributivist approaches to justice, exemplified in welfare state capitalist and socialist regimes, are best seen as secular strategies for bridging humanity's cognitive distance from God through collective action, specifically by alleviating, if not outright eliminating, suffering – and thereby possibly even completing creation. However, the strategy itself involves the recycling of evil as good, an 'end justifies the means' ethic in which both goods and harms are more equally distributed across society, in turn contributing to a sense of shared fate that otherwise might be lacking. In the third section, I explore the roots of this sensibility in the prophetic strain of the Abrahamic religions, which extols perseverance in the face of adversity. The sentiment remains even in Rev. Thomas Malthus' population theory, the intellectual antecedent of Darwin's theory of natural selection, in which all life forms – not least humans – are treated as rough drafts or experiments in the pursuit of some of higher order being that God struggles to shape out of recalcitrant matter. While Darwin famously lost his Christian faith because he could not accept such a callous deity (hence natural selection is nowadays taken to be, by definition, 'blind'), I argue that the moral horizons of a 'transhuman' being capable of reversing various currently irreversible features of the human condition may approximate those of Malthus' God. In the final section, I characterise this rather provocative prospect as 'moral entrepreneurship', which is already on its way to public acceptability in today's world.

1 Divine suffering and its human remediation, aka distributive justice

'Theodicy' is the discipline that deals with the terms of divine justice, or 'the ways of God as justified to man', as it used to be put. It is nowadays associated with a pale version of its former self, a boutique topic in the philosophical end of theology that questions how a perfect deity could have created such an imperfect world, especially one in which bad things routinely happen to good people. Although Max Weber (1963) made theodicy a central organising principle in his cross-cultural sociology of religion, 'theodicy' under the exact name began to exist only in 17th century Europe as a development in parallel with the rationalisation of the cosmos brought about by the Scientific Revolution. (Leibniz published the first book bearing the name 'theodicy' in 1710.) Once theologically inspired natural philosophers like Descartes and Newton proposed fundamental principles governing all of physical reality, the monstrously exceptional in nature started to appear as a tractable intellectual problem – as opposed to a fleeting glance into the mysteries of supernatural governance and possibly its interruption (whenever Satan was implicated). Instead of scapegoating something or someone contiguous with the monstrous event or simply praying to God for strength to survive the monstrous ordeal, theodicy demanded a more systemic approach that aimed at a deeper understanding, if not an outright improvement, of the divine order.

In this respect, 'evil' was raised from a criminal to a civil offence *vis-à-vis* cosmic governance. Natural disasters and heinous acts were no longer addressed in terms of the parties directly involved. Rather they were interpreted as part of an overarching design that transcends such local concerns but nevertheless is ultimately intelligible to those capable of adopting the right perspective, what Thomas Nagel (1986) dubbed 'the view from nowhere'. In the Newtonian cosmos, this was a position equidistant from all points in space and time, which is to say, a mathematical infinity. Although such a point was unreachable, we might still approximate it as a certain studied 'neutrality' towards the parties to a dispute. It would be difficult to exaggerate the significance that this 'geometrisation' of judgement has had in the modern era (Feyerabend 1999): Seeing things 'in perspective' now clearly suggested a check on our instinctive 'tit for tat' sense of right and wrong: that is, the routine doling out of pleasure and pain traditionally associated with commutative justice, which had been associated with 'natural morality' – in this case retribution – in both the Mosaic and the Aristotelian corpus. Without denying the need to identify the specific source of harm, theodicy treated that source, in the first instance, as a symptom rather than a target for the administration of justice.

For theodicists, a commutative principle such as 'an eye for an eye' constituted a primitive animal response biased towards restoring a prior sense of order indifferent to the injustices that had been normally tolerated in the absence of formal charges. Instead, theodicy demanded a more comprehensive mode of normative reflection, one that would result in more nuanced and perhaps even counter-intuitive judgements *vis-à-vis* the status quo. In light of such judgements, those other than the actual perpetrators may be required to redress the damage committed, as well as allowing those other than the actual victims to benefit from the redress. The most obvious modern juridical exemplars are laws associated with 'affirmative action' and 'positive discrimination', which administer group-based justice founded on a sense of causation that spans widely in space and time. The evolution of tax law also reflects this sensibility, as elaborated below. The overall image is one of converting individual punishment to a collective learning experience, whereby 'we' as equally invested members of a society come to a deeper understanding of the ultimate normative order, In modern secular society, this is typically expressed as some optimal mix of liberty and equality for its members. For the theodicist, to aspire to any less would be to betray our divine birthright, in this case our species-unique capacity to second-guess what God has in mind for us.

As soon as Leibniz baptised the discipline, theodicy's claims to 'second-guessing', and later 'playing', God were received with a mix of ridicule and consternation, as epitomised in Voltaire's 1759 novel *Candide*. Thus, Kant was well within *bien pensant* opinion when he argued in *Critique of Pure Reason* for the 'always already' failure of theodicy's cognitive feats, however necessary a belief in them may be for the conduct of our moral and intellectual lives. In effect, Kant hedged his bets, saying that even if God exists, by definition we could never acquire his exact perspective on things. At most, we might enact (or 'simulate') our own version of the deity's sense of universal legislation, as in Kant's own formulation of the moral law, the categorical imperative.

Yet even this half-hearted endorsement of our godlike abilities is indicative of the lasting impression that theodicy has left on secular normative thought. It comes out most clearly in the central debating point in the discipline's modern heir, political economy: *Is the unequal distribution of wealth in society deserved?* Theodicy's calling card is to suppose that a sensible answer might be given to this question. Here I mean to include both those who argue, *à la* Hayek, that any subsequent 'redistribution' is bound to make matters worse and those who argue, *à la* Keynes, that exactly such redistribution is required to achieve a just social order. The array of taxation, incentive and welfare schemes that followed in the wake of the latter response rendered measurable and calculable the Abrahamic idea that we are each our brother's keeper – and that all of humanity is debased whenever any human suffers unjustly (cf. Fleischacker 2004). Of course, this still left open questions about the limits of 'justifiable' individual suffering and the appropriate vehicles for alleviating or remedying such suffering. I shall return to this topic in the next section.

At this point, it is worth considering the subtle but deep hold that redistribution has had on the sociological imagination. After Parsons (1937), Emile Durkheim and Max Weber have been routinely portrayed as offering complementary foundations for the discipline of sociology. Usually these 'foundations' have been cast in ontological terms, as, say, collectivist *vs.* individualist, structure *vs.* agency or macro *vs.* micro in terms of their respective orientations to the study of social life. However, their most lasting legacies, which may outlive the discipline they mythically co-founded, relates to alternative accounts of social rationality, or *sociodicy*, in Jon Elster's (1983) evocative term. It has been common in the history of philosophy and politics to argue that reciprocal inequalities (aka complementarities) among people living together at the same time are necessary to hold society together. This thesis has been often expressed in terms of hierarchy as the universal social glue (Brown 1988). Durkheim and Weber distinctively shifted the focus to the unequal significance assigned to people living in different times, or *time discounting*, to borrow an expression from welfare economics (Price 1993).

On the one hand, people may 'discount the past' when they make it seem that what they are doing now is what they always have wanted to do. This strategy of 'adaptive preference formation', raised again in section 4 below, is perhaps most familiar from the Ministry of Truth in Orwell's *1984*, which continually rewrote the past to legitimise current state policy, but it is also implied in Durkheim's attempt at a unified theory of solidarity that could incorporate both 'primitive' and 'modern' societies through a generalised sense of 'division of labour' that would justify conformity in terms of prescribed social roles. On the other hand, people may 'discount the future' by using their past behaviour as the baseline on which to plot their subsequent actions, without considering the long-term negative consequences of such blind obedience to precedent, or what epistemologists call 'straight rule induction'. These take self-conscious form in the legal and economic practices that Weber associated with modern processes of 'rationalisation'. However, in the psychological literature, this form of time discounting is taken to transpire subconsciously through the acquisition of habits and addictions, whereby one comes to reproduce not the original action *per se* but its effects, which invariably entails increased effort over time (Ainslie 1992, 2001).

What justifies the Durkheimian and Weberian approaches to social time as constituting latter-day theodicies, or sociodicies, is their concern to capture the idea that, at any given moment, society has a fixed standpoint – as etymologically suggested by the word 'state' – in terms of which any event or action may be judged good or bad, progressive or regressive, etc. From this perspective, these approaches to social time may be understood as theories of the ongoing background work necessary for the construction of that godlike sense of a fixed normative order against the flow of day-to-day events. Combining Durkheim's and Weber's complementary approaches to time discounting, social order is maintained by past discounting compensating for the excesses of future discounting. For example, the main reason that prophecies of ecological doom never seem to be vindicated is not that they turn out to be false, but rather that different people live with the relevant consequences. This allows time for the new generation to want – or at least tolerate – what is most likely to happen, which enables shifts in how these predictions are measured and interpreted. In effect, the Durkheimian behaviour of parents enables their offspring to live as Weberians.

The state is strongly implicated in this process, especially in the production and maintenance of public records and a national education system. Indeed, the discipline of sociology arguably arose from a realisation of the state's innovative use of time, rather than space, as the organising principle for defining the boundaries of a society. (The relevant contrast here is with so-called traditional or primitive societies, in which social identity is circumscribed by its members' relatively limited physical mobility.) Thus, contrary to those who follow, say, Wallerstein (1996) in wanting to bury sociology in its 19th century nation-building roots, the phenomena associated with Western imperialism, World Communism and their ideologically neutered successor, 'globalisation', turn out to be consummately sociological, as non-Western peoples came to mark their own existences in terms of various Western time-keeping practices, ranging from 'living by the clock' to self-imposed world historic narratives of 'development'.

Turning to the missing member of sociology's founding Holy Trinity, Karl Marx, the failure of his doctrine is usually attributed to his underestimation of the state's role in counteracting the worst effects of capitalism so as to forestall indefinitely the rise of widespread revolutionary class consciousness. To be sure, the emergence of the welfare state in Germany at the end of the 19th century spelled the beginning of the end of Marxism's political efficacy. In this context, Marxism's failure is often cast as an object lesson in what Giddens (1976) calls the 'double hermeneutic' that allegedly distinguishes the human from the natural sciences – or more prosaically, that people can respond perversely to things said about them. Thus, once Bismarck learned of Marx's prediction that the global anti-capitalist revolution would start in Germany (given its level of organised labour), he encouraged legislative measures that launched the modern welfare state and effectively ensured that Marx's prediction would never be realised. But this interpretation of Marxism's demise, while correct as far as it goes, is no more than part of the story.

The sociologically more salient part is that the Marxist reliance on the labour theory of value was oblivious to the organisational role of time discounting in social relations. According to the labour theory of value, workers should receive a wage exactly proportional to the amount of labour they have invested. While this is recognisable as a principle of 'natural justice' from ancient and medieval economics, in the modern world there are very few, if any, social practices in which participants expect their costs to be met by proportional benefits. Indeed, this point is central to the state's role in rationalising social relations through mechanisms of 'redistribution' that take into account the way in which the perceived past and the anticipated future either enhances or diminishes the value of individuals' actual contributions to society – that is, Elster's sociodicy. Nevertheless, the labour theory of value remains useful as a theoretical baseline of 'society degree zero', in which individuals are valued exactly for what they themselves do – nothing more and nothing less.

In that case, we can understand the two general redistribution strategies in modern societies as opposing deviations from this baseline. Not surprisingly, both arose as direct responses to Marxism. The first, historically defended by the *fin de siècle* Austrian finance minister and founder of 20th century 'Austrian economics', Eugen Böhm-Bawerk

(1959), is associated with the Right; the second, defended by Cambridge's missing link between Alfred Marshall and John Maynard Keynes in the evolution of welfare economics, Arthur Pigou (2000), is associated with the Left.

Böhm-Bawerk argued that workers should receive less than what the labour theory of value prescribed because the employer takes the bulk of the risks by setting up work in the first place before any goods are sold. For him an 'exploited' worker is really the beneficiary of a loan that will be hopefully repaid to the employer by consumers. What Marxists demonise as 'surplus value' merely puts a negative spin on this repayment strategy. In contrast, Pigou argued that even non-workers should receive compensation at the expense of the wealthy, because the unemployed are not personally responsible for their idleness yet they are likely to do more than the wealthy with the redistributed income, for whom proportionally it matters less to their overall well-being. For Pigou the persistence of unemployment reflects neglected opportunities for capital investment, in response to which the state is entitled to penalise the wealthy through higher taxation. This allows the state to do with the wealthy's resources what they themselves have failed to do on their own. In effect, the wealthy are being punished less for selfishness than lack of imagination – that is, in the words of that rogue Austrian, Joseph Schumpeter, their failure to 'creatively destroy' their market advantage.

In short, Böhm-Bawerk discounted the future in favour of the past, Pigou discounted in reverse. Each perspective incurred the wrath of the other throughout the 20th century, during which they defined the limits of the social policy imagination. At the start of the 21st century, Böhm-Bawerk appears to have the upper hand, as states are now more inclined to provide incentives to start up new businesses (and allow them to retain much of the resulting profit) than to penalise wealthy people who fail to be entrepreneurial or charitable (by forcing them to forfeit some of their wealth in the name of collective welfare). Marxism as the putative 'no discount zone' failed because, in the end, the labour theory of value provided at most a regulative ideal towards which an existing society might aim but little guidance in either identifying or redressing 'differences that make a difference' in such a society. In this respect, Marxism lacked a distinctly sociological sensibility, which is perhaps most evident in the movement's eschatological impulse to end history formally and create society anew. Theologically speaking, if one holds a view as perfectionist as Marxism, then given the world's persistent flaws, you will be inclined either to anticipate an impending afterlife of vindicated souls or, more secularly, to redress the injustices oneself by a revolution that installs new people informed with a new ideology – the existential horizons that Voegelin (1952) stigmatised as 'gnostic'.

In light of such politically extreme solutions, the tax system looks like a relatively reasonable vehicle for administering justice across classes and generations. This helps to explain the Marxist attraction to social democratic parties, even if such parties are liable to treat the sheer maintenance of parliamentary power as an end in itself (Michels 1959). Taxation facilitates the incorporation of the private into the public sphere by targeting people not by their words but their deeds, specifically, income and expenditure. This places the tax official in a position to judge whether people have made the best use of the resources allotted to them and, if not, to redistribute accordingly. Thus, arguments for the morally advanced standing of welfare states have relied on their willingness to convert criminal offences to tax-like civil offences. In this context, the harming of an individual life is treated primarily as the deprival of societal resources, which is best redeemed not by indefinite incarceration but some appropriate compensation, often in the form of labour but increasingly money. Indeed, some legal theorists today outright praise money as a medium of universal exchange that permits very creative ways of righting both public and private wrongs (Ripstein 2007). However, the spirit of this proposal needs to be seen correctly. Such a homogenisation is favoured not because all acts, events and things are intrinsically of similar value but because they are all traceable to a human source – another case of human labour simulating divine power in matters of value creation.

2 The human simulation of divine legislation, aka justified suffering

The phenomenon of time discounting also draws attention to a perverse sense in which human and divine legislation may converge: Those who construct the norms rarely live with the most important consequences of their application. Indeed, except in explicitly constitutional matters, lawgivers tend to be relatively immune to the norms they design; rather, offspring, immigrants and even unsuspecting third parties are the main downstream recipients of the norms' effects. At the empirical level, this already suggests that sociologists always need to ask who are likely to be held accountable to publicly advertised social norms. At a theoretical level, it explains the sterility of the structure-agency debate that dominated social theory in the second half of the 20th century: Once time discounting is given its due, it becomes clear that we are born into structure and beget in agency (Fuller 1998a; Fuller 1998b). Put another way, we are most free when the consequences of our actions do not affect us one way or another.

This point pertains to the conduct of sociological research itself. 'Reflexive' moves in sociology tend to be destabilising because they demand that the researcher sees herself as fully embedded in the causal nexus – that is, not simply as enacting an instance of established role-expectations. Reflexivity may be an intellectual virtue or a political device, but it is *not* a regular feature of social life. Indeed, much deliberate effort goes into suppressing the emergence of reflexivity at a societal level, which explains relative ease with which the baselines of official statistics and the content of the national curriculum can change. If students had to learn not only what their teachers teach them but also what their teachers were taught, the discrepancy between 'ought' and 'is' would be a source of endless cognitive dissonance. But 'luckily' (at least, and perhaps only, from the standpoint of social stability), students only learn what their teachers think they should learn, which through a combination of faulty memory and judicious editing the teachers can easily pass off as what they themselves had been taught. Thus, with an ironic nod to Giddens (1984), what both teachers and students tend to believe is being transacted in the name of 'structuration' in fact constitutes the terms on which 'agency' is intergenerationally reproduced.

Such a lack of reflexivity is canonised in the classical sociological literature as the distinction between 'substantive' and 'functional' rationality, i.e. the difference between rationalising an activity's end and rationalising the means by which it is achieved. From Max Weber onward, it has been common to view the cognitively diminished (aka unreflexive) state represented by functional rationality – whereby the ends are simply taken for granted – as a high but perhaps fair price to pay for the efficiency of our modern ways. However, the 21st century is already revisiting issues relating to substantive rationality because various ends that the 20th century had presumed to be jointly pursuable may soon require serious trade-offs. There is nothing like scarcity to focus the mind on value commitments, which amounts to a wake up call to reflexivity. Heralding this shift from means to ends have been the various ecological crises that the planet has been facing as a by-product of 'overdevelopment' since the 1970s. At the extreme end of these crises lies the problem of how to decide which from among various humans and non-humans to include in the social order, given the emergence of normative intuitions that favour the promotion of 'healthy' non-humans over 'unhealthy' humans (Fuller 2006b: Part Three).

A striking feature of putting the matter this way, which owes much to the persistent efforts of the philosopher of 'animal liberation', Peter Singer (1975), is its unholy alliance of Aristotle's eudaemonism and Bentham's utilitarianism. To be sure, both are consequentialist moral systems – that is, the good is judged in terms of the ends that are served. In the former case, each type of being, or species, has its proper sense of self-realisation, which a 'good life' aims to achieve. In the latter, one is attuned to the global ratio of pleasure to pain, with the stress placed on the minimisation pain rather than the maximisation of pleasure. The overall result is a *precautionary bioliberalism*, in which societal membership is limited only to those individuals who can jointly pursue fulfilling lives as conveniently as possible (cf. Fuller 2006b). Thus, the parameters of policy are defined by pre-empting the suffering of the unwanted – be they unborn or outlived – as well as those otherwise flourishing individuals who would become 'second-hand sufferers' by having to fund or attend to the first-hand suffering of the unwanted.

Whatever else one might say about precautionary bioliberalism, it is an increasingly powerful secular theodicy, or sociodicy, that, on the one hand, does not restrict membership in society to *Homo sapiens* yet, on the other hand, within our own species includes only those individuals capable of a naturally free existence. In the politically correct euphemisms of actor-network theory, such individuals are nodes that help to extend already existing networks rather than provide barriers to their extension (cf. Latour 1987). It is worth recalling that actor-network theory's patron saint, Durkheim's great nemesis Gabriel Tarde, was originally inspired by the social behaviour of *ants*, whose pattern of environmentally constrained, path-dependent imitation would a century later move the Harvard ant specialist E.O. Wilson (1975) to propose 'sociobiology' as the completion of the Neo-Darwinian synthesis. Indeed, Tarde's fundamental mimetic principle – that innovation results once imperfect imitation passes for an invention thereby setting a precedent for others to follow – was borrowed from the landmark account of the ant colony as an 'animal society' provided by Herbert Spencer's French translator, Alfred Espinas, which differed from Mandeville's swarm of bees allegory that had provided a naturalistic underpinning for Adam Smith's account of the division of labour a century earlier.

What is so distinctive about *ants* as social creatures that would inspire such grand visions of social science? In brief, ants constitute a species whose members' collective efforts amount to a mass incorporation of their colonised environment. The shape of colonisation is largely determined by the first successful adaptive response to the target environment, whose subsequent changes are met by equally successful adaptive responses on the part of the ant society, even at the cost of particular individuals, who may figure, say, as food. This makes ants amongst the most difficult species to eliminate once they have found a home, since they appear predisposed to find the path of least resistance to collective survival. In this respect, an interesting missing link between Tarde and Wilson is the latter's Harvard inspiration, William Morton Wheeler. Wheeler's 1911 article 'The Ant Colony as Organism' suggested (admittedly without a modern understanding of genetics) that the members of an ant colony merged into an effective 'superorganism', which may have provided a primitive evolutionary vehicle for what philosophers of the day were calling 'collective consciousness' (Parikka 2008).

While the ant's eye-view of the world certainly counts as a sociodicy and might even serve as the basis for a coherent theodicy, it would not be one compatible with the classical version of theodicy that Leibniz named and Voltaire subsequently ridiculed. Theodicy's alleged achievement of a neutral juridical horizon – the 'view from nowhere' – assumed that humans, having been created (as Genesis says) in 'the image and likeness of God', differ in principle from the deity by degree rather than kind. After Augustine's exegetical emphasis, it has been called the *imago dei* doctrine (Fuller 2008a: chap. 2). In other words, even if God is infinitely more wise or intelligent than we are, the difference constitutes a continuum. As medieval scholastics following the Franciscan John Duns Scotus put it, 'wise' and 'intelligent' are *univocally* predicated of humans and God: that is, the words mean the same when talking about the two beings. One consequence of this semantic reductionism, or literalism, is that whatever role God assigns to humanity in his cosmic plan, it must be, like the deity's plan itself, potentially subject to self-legislation. Bluntly put, to be accorded the respect to which we are entitled by virtue of having been created *in imago dei*, we must be able and allowed to choose to be part of the divine scheme, as if we had a hand in its design. Whatever other social virtues ants might enjoy, self-legislation is not one of them.

In that case, Tarde's and Wilson's entomological sociodicies fail by the standards of classical theodicy because they depend on a collection of ants merging with the environment to produce a superorganism whose survival rests on its members being treated as mere parts rather than autonomous agents. In contrast, the Abrahamic tradition of theodicy provides only two ways to characterise the God-human relationship. In each case, unlike the ants, we have free will. However, in one case we decide to learn from nature to conform to the divine plan, while in the other we decide to take nature's incompleteness as an opportunity to improve if not outright complete it in accordance with the divine plan (Fuller 2010: chap. 7). The two scenarios correspond to the ones defended by Leibniz and his Cartesian rival, Nicolas Malebranche, respectively, in theodicy's original late 17th century flowering (Nadler 2008). To be sure, both scenarios allow for the possibility of failure, but both would have been the product of our own decisions, the consequences of which we can anticipate only to a limited extent.

Physics was another naturalistic source for sociodicy. The US founding fathers famously designed the constitution of the new republic on the model of Newton's mechanical world-system, with its separation of powers and checks and balances (Cohen 1995). However, Newton's universe was still one in which all individuals are interchangeable and governed by timeless laws. A subtler basis for sociodicy occurs once physics incorporates statistical thermodynamic principles in 1870s, largely following a stirring public address by James Clerk Maxwell, more about which in the next section. As in the case of biology, the lure of statistics here was the prospect of acknowledging – as the Newtonian worldview had not – variation amongst individuals without needing to understand the mental states of each individual in order to discern emergent tendencies in a population. However, in the social physics derived from thermodynamics, the individual is treated not as a part in a whole but an instance in a tendency.

Yet neither the physics- nor the biology-based sociodicy regards the individual as an autonomous agent. At least, so it seemed for most of the 20th century, when naturalistic theodicies remained rather alien from commonsense understandings of the world – let alone classical theodicies. However, I shall return to this point because as denizens of the 21st century get used to the idea that they are 'always already' risking their individual lives in various ways, the consequences of which potentially offer lessons for future generations, it is reasonable to suppose that these very naturalistic sociodicies are coming to be internalised as part of our self-understandings, and in that sense 'reflexively' applied. In effect, we have come to accept that there is a 'statistical' aspect to our being-in-the-world. In that case, again from the standpoint of reflexivity, social scientists who continue to champion a strong 'qualitative' *vs.*

'quantitative' distinction in research methodology may be guilty of perpetuating of a conceptually artificial dichotomy – perhaps no more than a Neo-Kantian atavism – that fails to do justice to social agents who have already arrived at ways of blending the two perspectives. This is arguably the deepest lesson that science and technology studies can teach the social sciences more generally (e.g. Latour 1987).

But I shall round out this section by sticking to 20th century sensibilities. Perhaps unsurprisingly, within a half-century of the introduction of the thermodynamically-oriented social physics, the founder of the Harvard sociology department, Pitirim Sorokin (1928) comprehensively dismissed it as fatalism recast in mathematics. For example, social physics attempted to rationalise violent social revolutions as akin to the 'critical phase transitions' that a fluid undergoes to become a gas. Treating masses of people in the same way as atoms leaves the impression that that their collective mood will eventually change from inert solidity to liberal fluidity and, finally, gaseous revolt. No further explanation is needed: the motives and thoughts of the individuals involved are irrelevant.

Lest one mistakenly assume that Sorokin had the final word, one need only turn to Phillip Ball, a science journalist trained in chemistry, who won the 2005 Aventis Prize – Europe's leading popular science award – for *Critical Mass*, a book that aims to show 'how much we can understand about human behaviour when we cease to try to predict and analyse the behaviour of individuals and instead look to the impact of hundreds, thousands or millions of individual human decisions' (Ball 2004). Ball's main examples are traffic flow, coalition formation and marriage selection. What these rather diverse social phenomena share is that when scientists observe them they assume that they see exactly the same set of conditions as the real people involved: only so many routes for a driver to choose from when reaching a particular junction, only so many parties with which one can form a government and only so many partners to choose from when deciding who to marry. Of course, social scientists, probably more so now than in Sorokin's day, find this a huge oversimplification that obscures the real human condition. For example, when a person decides who to marry, he or she considers many factors, including potential future income, quality of life and happiness. In fact, many decisions about many aspects of life are being made simultaneously, issuing in multiple causal chains. Even if the final decision can be understood as an instance of an overarching statistical tendency, how it came to fall under that particular tendency may be quite unlike the path taken by other decisions that also end up contributing to the same tendency.

Nevertheless, the resurgence of naturalistic sociodicies reflects a larger trajectory, only partly orchestrated, to reabsorb the social into the natural sciences, whereby 'human nature' becomes a proper subset of nature. I have characterised this trajectory as the 'casualisation of the human condition' (Fuller 2006b). At the empirical level, studies broadly associated with 'evolutionary psychology' show that non-humans appear to possess qualities such as consciousness, meaningfulness and even altruism that were classically unique to humans; at the normative level, advances in medical technology has effectively demystified the 'sacredness' of individual existence by rationalising practices such as abortion and euthanasia that ease the passage in and out of the human condition. In the one case the distinction between human and non-human is blurred, in the other it becomes easier to think of life in locally functional terms. To preserve human dignity in these times, such casualisation needs to be tempered by a sense of individual autonomy that enables one to take full responsibility for one's own fate, regardless of its statistical predictability or social utility. An important concept in this context is 'justifiable suffering', to which we now turn.

3 How might suffering be justified in the days to come?

The previous two sections highlighted a persistent secular version of the problem of theodicy in social science, theory and policy: How much suffering is tolerable in order to redistribute resources to produce a just society? Generally speaking, individual suffering as such has not been seen as a break point in modern ethics. On the contrary, modern ethics arguably presupposes a *metatheory of justifiable suffering,* namely, that the moral agent would be willing to suffer, if she had a full understanding of the ends served by her suffering. Such judgements are bound up with an ideal self-understanding, which in turn implies knowledge of the groups in terms of which one's identity is defined. Justifiable suffering is most evident in a utilitarian calculus capable of translating individual experiences of pain into publicly relevant costs. Ideally the sufferers themselves would incorporate these translations into their self-understandings, perhaps as grounds for political mobilisation against the corresponding injustice, once a certain threshold of pain has been passed. In that case, one would risk still more suffering – say, in the form of violent revolution – in the hope of eliminating its cause.

Suffering's rationally mediated impact on the modern moral imagination applies even more strongly to utilitarianism's great rival,

Kantianism. Begin by recalling Kant's (in)famous claim that one should always uphold the truth, even if it endangers the lives of loved ones. It was made partly in response to the French liberal Benjamin Constant, who equated inhumanity with suffering. For Kant such an equation, despite its surface attraction, only served to reduce the human to the animal (Rosen 2008). In contrast, our humanity is upheld only when we are treated as pure ends, which is to say, beings capable of dictating the terms of their own existence, as opposed to a helpless bundle of passions seeking the optimal balance of pleasure and pain. As Kant made quite clear in his 1784 essay 'What is Enlightenment', we are most demeaned when we are treated like children or, worse still, pets. In both cases, our primary attitude is one of protection rather than empowerment. In contrast, the Enlightenment mandate is that we strive to produce a world that tolerates error, so we may learn from it – not one that avoids error, so that we may be spared its accompanying pain. The policy upshot is that suffering is inevitable and perhaps even permissible – as long as it is registered publicly, so as to inform further (self-)legislation. This helps to explain the kinds of politics inspired by both Kantianism and utilitarianism.

On the one hand, Kantianism has been repeatedly invoked to justify self-sacrificing patriotism based on one's self-identification as a 'citizen' who, by that definition, is obliged to do whatever it takes to promote the national interest, even if it means fighting in wars in which one had no role in starting and no wish in perpetuating. However, the prospect of personal death is not senseless but rather the ultimate affirmation of an ideal pursued for its own sake whose full realisation may be enjoyed in the future by other citizens inspired by their example. Put more bluntly, a Kantian may be the ideal means to a utilitarian's ends. In this context, the revolutionary spirit of the past – as disseminated through authorised national histories – functions as citizenship's genetic code, which is fully realised only by facing the new challenges thrown up by the political environment. Thomas Jefferson even suggested, as a safeguard against constitutional complacency in the new American republic, that it might be a sign of political health if a violent revolution broke out as each new generation came to maturity. (Let us set aside the tricky problem of exactly demarcating such 'generations', given that new people are always being born.) Thus, self-sacrifice for the revolutionary cause satisfies Kantian scruples as an action done for its own sake, while under a utilitarian guise appearing as a more or less adequate means of achieving the ends that the revolutionary claimed to uphold.

On the other hand, utilitarianism has tended to put a more biological spin on the same ideal, in line with the populational worldview introduced by Thomas Malthus and generalised by Charles Darwin. Accordingly, the value of each individual is as a living experiment in the struggle for survival. The 'struggle' is a negative by-product of individuals pursuing their own interests in a material environment fundamentally indifferent to their success. In other words, opposition to self-realisation comes only indirectly from other would-be self-realisers but directly from a external world subject to scarcity. This point helps to explain the appeal of David Ricardo's brand of political economy in Marx's formative period: namely, one might increase the level of general welfare by increasing productivity, which in turn requires the replacement of finite natural resources with indefinitely extendable artificial ones. In that sense, we bend the world to our will. Thus, while human labour may be normally self-expending, that labour-power may be captured and embodied more efficiently in machines. Though not Ricardo's concern, consumer goods arguably have also taken the same trajectory, as the potentially scarce agricultural resources traditionally needed to sustain life have morphed into a cornucopia of synthetic commodities whose replacement is dictated less by its supply than by a demand that may be itself manufactured.

The problem here, as Marx saw very clearly, is that the external world's recalcitrance is simply shifted elsewhere, as greater general welfare is rendered compatible with more immiserated individuals. Moreover, the sense of 'immiseration' is doubled under capitalism. Not only is the number of people living in economic misery increased but even those not living objectively miserable lives remain subjectively miserable because they feel forced to compete in various markets for 'positional goods', aka 'Keeping up with the Joneses' (Hirsch 1976). Nevertheless, capitalism's robustness in the face of 200 years of social, political and economic boom and bust cycles speaks to a peculiar synthesis of the Kantian and utilitarian perspectives – what Joseph Schumpeter (1961) identified as the entrepreneur's ongoing 'creative destruction' of markets. The impulse of the entrepreneur (or 'venture capitalist', as we say today) is *not* to stick indefinitely with a product that he has brought to market, even if its rate of return on investment is high. In that sense, he is motivated by neither the utilitarian's cost-benefit considerations nor the Kantian's faith in the product's intrinsic worth. Rather, the entrepreneur adopts the second-order perspective that values efficiency for its own sake – that is, the ultimate utilitarian virtue treated in a Kantian manner. Thus, the entrepreneur is always thinking about ways to supersede the product at hand by inventing one that either produces the same effects at a lower cost (to the producer, though perhaps not to the consumer) or induces the consumer to re-define, aka 'upgrade', her wants such that the new product satisfies those better.

To be sure, Galbraith's (1958) phrase 'planned obsolescence' has put a cynical spin on this process, which nowadays has been extended to the realm of knowledge production itself, most noticeably through increasing scientific specialisation (Fuller and Collier 2004: chap. 7). However, it is also arguably a way to simulate divine creation in concrete human terms, especially if one takes seriously that even Genesis – a book not normally read as a defence of evolutionary theory – portrays God as having created in stages over several 'days', throughout which the deity drew feedback and improved output to better match his original plan. Lest this account seem fanciful, it clearly influenced the Unitarian-raised mathematician Norbert Wiener, father of cybernetics and author of the National Book Award-winning *God and Golem, Inc.* (1964).

This book, which expressly casts the quest for 'artificial intelligence' as a high-tech version of the *imago dei* doctrine, launched the 'cybertheology' subsequently advanced by, among others, William Sims Bainbridge (2005), who is best known to sociologists as the co-founder of the rational choice theory of religion, which treats the emergence of Christian religious denominations as a species of market differentiation (Stark and Bainbridge 1996). However, as we saw in Chapter 3, in his day job as head of social informatics at the US National Science Foundation, Bainbridge co-authored perhaps the most influential science policy statement of the past decade, which effectively launched the drive to focus the emerging nano-, bio-, info- and cogno-sciences and technologies on research designed to 'enhance human performance' by overcoming the carbon-based barriers (e.g. ageing, fatigue, limited storage capacity to a full realisation of our potential (Roco and Bainbridge 2002a and 2002b).

The road from theodicy through political economy and cybernetics has perhaps come full circle in the person of Craig Venter, the venture capitalist whose patented automated technique for sequencing genomes has significantly closed the gap between the knowledge management metaphor of 'data mining' and the more literal idea of mining raw materials. Courtesy of intellectual property legislation, 'biocapital' has been effectively assimilated into the primary sector of the economy (Fuller 2002a: chaps. 1–2). Without denying its Marxist and Foucaultian resonances (e.g. Rose 2006), this line of thought has firm roots in molecular biology's historical self-understanding. Already 20 years ago Harvard's chair of molecular biology, Walter Gilbert (1991), himself a Nobel Prize winner, echoed the then half-century-old prediction of Erwin Schrödinger (1955) that his discipline would become a branch of informatics, in which teams of biologists would selectively exploit molecular combinations in search of new biochemical resources, perhaps of medicinal value or even new synthetic life-forms.

It is worth noting that all parties to this line of thought – and policy – have stressed its 'trial-and-error' character. In this important respect, Venter and Gilbert are one with Malthus and Ricardo in upholding the disposability of living matter. The difference is that the former pair have advocated it for beings below, so to speak, the 'moral radar' of personhood on which natural law theorists and their secular socialist descendants have had their sights trained. However, once we envisage a continuum between selecting particular complex molecules and selecting entire organisms and populations, we are on the slippery slope that is theodicy's stock and trade. In short, we play God. From this standpoint, what happens to get called 'divine' or 'natural' or 'artificial' *vis-à-vis* the selection of living matter is more affective than substantive. While we may wish to live in a world that deploys this three-way distinction in a discriminating fashion, the three predicates are literally referring to the same process. For example, it is common for conservative defenders of natural law to be vehemently against abortion and murder yet with equal vehemence uphold capital punishment, even though all three acts are instances of negative selection for some putatively positive purpose.

My guess is that with the normalisation of, on the one hand, stem cell research (which breeds and farms embryonic cells to grow functioning organs) and, on the other, euthanasia and organ donation, it will become harder to defend the discriminating appeal to selection associated with what I have just identified as the 'conservative' position. This is because the locus of legal concern will shift from the acts themselves to the consent of the individuals involved – where consent to self-destruction is taken by all ideologies to be, at least in principle, a legitimate option. It is telling that today's debates surrounding euthanasia increasingly centre on whether the patient has been presented with all available courses of action – not simply the ones that 'the loved ones' would find convenient. But once that array has been presented, and assuming that the patient is of sound mind, her decision stands. We thus appear to be living in a time when scientific progress is clearly forcing everyone *malgré lui* to become politically more liberal.

The only way to prevent such 'creeping liberalism' would be to prohibit outright certain forms of biomedical research, which in turn would arguably entail a much greater violation of our humanity than whatever 'individuals' might be sacrificed in a self-legislated selection process (Fuller 2010: chap. 1).

Once again, the Biblical precedent is relevant, since God appears to have no problem superseding and even eliminating entire populations of organisms, so to retain whatever is good about them despite their imperfect material embodiments. Indeed, just as a 'literal' reading of the story of Noah in the 18th century enabled the discovery of rock layers beneath the earth's surface to be interpreted as indicative of successive catastrophes, a 'literal' reading in the 21st century might easily focus on the imperative to save specimens rather than populations of organisms, in which case the drive to sequence genomes of various species for purposes of preservation should take precedence over the simple preservation of already living organisms. Thus, Noah's ark becomes the friendly laboratory capable of surviving a hostile ecology. Implied here is a specific understanding of 'divine omnipotence' that in turn serves as a benchmark of human performance: God ultimately gets his way but perhaps by somewhat indirect means in the face of a less than compliant nature.

It was in just this spirit that the great theorist of electromagnetism and non-conformist Christian, James Clerk Maxwell, first proposed that physics adopt a statistical approach in a famous 1873 address to the British Association for the Advancement of Science. Here he introduced the idea that a physical system such as a gas or liquid can be explained in terms of the average behaviour of large numbers of molecules, a task that did not involve what Maxwell dubbed the 'historical' task of tracking down the state of each individual molecule. Maxwell had been inspired by perhaps the most impressive instance of 19th century applied social science, the UK national census, which divided and correlated samples of the British populace according to age, tax payment, education, religion and criminal conviction (Porter 1986). Maxwell proved persuasive. Thermodynamics and optics were soon recast in statistical terms – and to great effect, as the latter provided the basis for quantum mechanics. Maxwell had regarded these social statisticians as the scientific vanguard of his day because of the efficiency with which they wielded their mathematical skills, which he thought may also provide a glimpse into divine governance, namely, that God treats individuals as literal experiments in living that may fail but whose failure is never meaningless, by virtue of the information such a life would have provided for those responsible for realising the divine plan.

In the same spirit, we also need to revisit the biology-based sociodicies. To take Genesis literally is to concede that reunion with God after Adam's Fall will be a long, arduous process that is compounded by our relatively long individual lives, at least when compared with other species. But for precisely that reason, ever since Thomas Hunt Morgan's introduction of the fruit fly to the experimental zoology laboratory at Columbia University in 1910 (which led to the discovery of the chromosome), the shorter life expectancies and hence quicker intergenerational turnover of these species have served as bionic models – or 'model organisms', in the normatively inflected terms used by biologists – for features and processes that we might like to regulate throughout nature, including our own bodies. Extended beyond the lab, 'biomimetics' developed as a field of engineering oriented to treating organisms as technological prototypes, which (as noted in Chapter 2) unsurprisingly happens to be the field of research of the leading British intelligent design theorist. To be sure, it is possible to see this development in purely pagan terms as a rigorous update of the ancient Greek fabulist Aesop, who observed in the normal lifespan of animals lessons that otherwise would have taken generations for humans to learn from studying only their own species. But given the ease with which genetics has been historically aligned with eugenics – virtually every major geneticist has been a 'eugenicist' of some sort (Pichot 2009) – this 'learning from animals' is more in the spirit of simulating the divine standpoint by collapsing centuries into hours, thereby enabling us to acquire some of the emotional detachment that theodicy's defenders – not least Malthus – ascribed to the deity. In any case, model organisms brought the discipline of genetics closer in spirit to its etymological roots in 'Genesis'.

Moreover, once equipped with nature-inspired technologies, we might think more imaginatively (aka divinely) about the terms on which 'the greatest good' can be secured for 'the greatest number', especially how *parts* of individuals might be subsumed under this rubric. Steven Lukes (1996) satirised the prospect in a society called 'Utilitaria', whose motto was 'From welfare to farewell'. Such 'Utilitarians' understood their life cycle to proceed from a childhood in which others provided for them, through their own self-provision as adults, to finally their provision of others in old age, as first their wisdom and then their organs are transferred to the young: One begins life as pure consumers and ends as purely consumed. This sense of one's personal history receives widespread endorsement because the cannibalisation of first one's experience and then one's very matter is understood as a form of self-transcendence – that is, one lives on in the bodies of successive generations. In that respect, the policy may be seen as moved by a sentiment rather similar to eugenics – namely, to rationalise inheritance – but in a way that is not politically coerced but self-legislated.

Given the extremes to which distributive justice is prone, it should come as no surprise that in either its Kantian or utilitarian forms, distributivism has had its trenchant critics. In particular, the peculiar coalition of free market thinking and Ultramontanist Catholicism (e.g. Joseph de Maistre) that first opposed redistribution in the late 18th and early 19th centuries remains very much alive today in the Austrian school of economics that descended from Böhm-Bawerk, crystallising around Ludwig von Mises in 1920s Vienna, whose members included Friedrich von Hayek and Alfred Schutz (cf. Hayek 1952; Camcastle 2005). The Austrians were notoriously sceptical about quantifying costs and benefits beyond the prices that people pay in specific transactions. Utility for them was purely subjective, as the price mechanism produced no more than snapshots of an ongoing process that could neither be predicted nor controlled. Understood charitably, the Austrians were celebrating the inviolability of human freedom and diversity. But truer to the implicit metaphysics of the situation, it amounted to a barrier placed between human and divine understandings of the human condition. After all, if all humans are indeed equal under the eyes of God, then a common standard for judgement and policy should be possible – and is, when left to the deity. But there's the rub: The Austrians hold that our divine image differs from God's original being in *kind* rather than degree, which renders the epistemic significance of prices beyond the original transactions a theological mystery. While human souls are transparent to their divine creator, they remain opaque to one another.

Nowadays bolstered by evolutionists who use Neo-Darwinian arguments to draw Spencerian *laissez faire* conclusions, the Austrians hold that any second-order tinkering with the natural spread of goods and harms in a self-organising population is bound to leave its members – and quite possibly their offspring – worse. As I have suggested, this fear goes beyond the mere admission of epistemic error to implicate sacrilege in the pretence to such godlike second-order powers. Albert Hirschman (1991) has aptly dubbed this sanctimonious sensibility 'reactionary', since it confers overriding normative significance on such past-oriented qualities as the durability and tolerability of living arrangements, regardless of whether they enable everyone to realise their full potential, either individually or collectively. From this standpoint, government is simply about ensuring that people are not forced into fates worse than the ones they would otherwise have most likely had. The profound pessimism of such a worldview is the secular descendant of Original Sin, in which human nature is portrayed as incorrigibly fallible.

While the redistributivist mentality retains from theodicy a sense of the pervasiveness of imperfection – indeed, evil – in both nature and morals, it grants humans the capacity for correction, improvement if not ultimately perfection. Thus, suffering is *not* to be necessarily avoided or immediately eliminated. Rather it is treated in more studied and forensic terms, as a sign from which society may emerge stronger than before, even if at the cost of particular individual lives. Put this way, redistributivism appears very cold, perhaps inhumane. Yet, such pretence to Olympian neutrality was made theologically available to all humans by the *imago dei* doctrine. Thus, according to theodicy's secular offspring, utilitarianism and Kantianism, a moral decision taken for oneself would be agreed by all as the best to have been taken, either because they would have weighed the utilities of the decision similarly (Bentham) or they would have applied the same decision to themselves (Kant). For their part, redistributivism's reactionary opponents regard all humans as equal *not* in their common descent but their common subordination to God. Contrary to the Kantian ideal of Enlightenment as humanity's release from intellectual and political nonage, the reactionaries deny that the parental nature of 'God the Father' is ever outgrown in some process of spiritual maturation.

At stake between the redistributivists and the reactionaries, one might say, is the *locus of re-enchantment – humanity or nature?* Does the divine reside within us struggling to come out or is it already present in a world that waits for us to find our place? In the former case, suffering is the by-product of a spirit trying to transcend the limits of its material container; in the latter, it signifies a being that has yet to adapt its beliefs and desires to its circumstances. Consider the ongoing policy concern with the planet's ecological de-stabilisation, which 40 years ago was attributed specifically to 'overpopulation' but nowadays, in more politically correct terms, to 'increased carbon emissions'. Is this to be resolved by a 'technological fix', whereby we manage to beat thermodynamics at its own game by arriving at more energy efficient means of achieving the same ends, say, via nuclear power? Or will it be solved by downsizing the human condition so that in the future there are not so many of us wanting so much? Although both options are recognisably secular and political, they are fraught with theological significance. The former vindicates monotheism (i.e. the religions of an enchanted humanity), the latter pantheism (i.e. the religions of an enchanted nature) – the source of the great West-East divide in religious thought that has existed for at least the last 150 years.

That the supreme deity has its own sense of justice is unquestioned by all believers. But is our access to it sufficient to repay systematic study – as opposed to the occasional gnomic remark? Max Weber's (1963) attribution of a 'theodicy' to each of the great world religions was originally a generous gesture that in terms of today's cultural politics may uncharitably appear 'ethnocentric', since only the Abrahamic religions – and especially Christianity – have pursued the matter in such a bloody-minded way that conjured first images and then formulas of God operating with a balance sheet that humans need to discharge in some publicly accountable fashion. Even latter-day boosters of pantheism, such as Peter Singer, despite his attraction to Buddhism, nevertheless remains enough in the Abrahamic orbit to conceive of a global welfare function that ranges across all species in its concern for the distribution costs and benefits (Singer 1994). Unlike the Eastern pantheists, who have simply proposed withdrawal from secular affairs as an antidote to the world's ills, Singer's various schemes for imposing a global tax on the rich and reconstituting health care systems to subsume all sentient beings speak to a lingering sense of human responsibility, though the original theological sense of privilege has long disappeared (Singer 2004).

4 Can either wisdom or prophecy justify suffering in a secular age?

At this point, it is worth recalling a popular late 19th century way of distinguishing the Western and Eastern (more precisely, Abrahamic and non-Abrahamic) world religions that would have been familiar to Weber – namely, between faiths based on *prophecy* and on *wisdom* (Masuzawa 2005). The former take a more active and prospective view of the world, the latter more passive and retrospective. This conceptual divide projects a massive difference in moral psychology and, more generally, human affect, which in turn influences the terms in which theodicy might be cast. Both faith positions are symptomatic of what Hegel called the *unhappy consciousness*, that is, the artificial separation of thought and action, mind and matter, or in disciplinary terms, philosophy and politics. However, the 'unhappiness' is expressed rather differently. The value that wisdom places on hindsight is associated with the cultivation of guilt and regret for actions that perhaps should not have been done and, in any case, cannot be undone. Here thought trails action. In contrast, the value that prophecy places on foresight is associated with the cultivation of hope and anticipation for actions that ought to be done and may well be doable. Here action trails thought but aims to catch up.

It is telling that even though the original Greek *philosophia* is normally translated as 'love of wisdom', the Western philosophical tradition has been relatively stingy in its display of such affection. Indeed, this long-standing suspicion is reflected in the resistance that the philosopher of science Nicholas Maxwell's (2007) faced when he first called for science to move 'from knowledge to wisdom' a quarter-century ago. Indeed, the pattern of response recalls what politicians make of calls to 'return democracy to the people'. Given what democracy is *supposed* to mean, who could *openly* disagree? Thus, rather than saying that the argument is fundamentally wrongheaded, a few earnest critics quibble with its details, while the vast majority pass over it in stony silence. Actual philosophers despise 'philosophy' just as actual democrats despise 'democracy'. However, the inconsistency in both cases is due less to hypocrisy than catachresis. In other words, 'philosophy' does not quite capture what normatively acceptable histories of Western philosophy are about, just as 'democracy' does not quite capture what normatively acceptable histories of democracies are about. But more importantly for our purposes, not only has Western philosophy never really been about wisdom.

Consider the following inventory of generalisations about what 'wisdom' has meant in the Western philosophical tradition. Among contemporary philosophers, the Neo-Thomists would probably most warmly embrace the sentiments embodied in this list:

- Wisdom is possible in both action and thought. The former, also known as 'prudence', consists in finding a middle way between extreme courses. The latter, also known as 'contemplation', consists in coming to a state of equanimity.
- The individual grows in wisdom but society does not: Everyone must become wise for themselves.
- There is 'modern science' but no 'modern wisdom'.
- Science, but not wisdom, can be 'misused'.

- Wisdom has an integrative function within the individual person, whereas science distributes its inquiries across many specialised individuals.

An interesting feature of modern philosophy, starting with Descartes and Hobbes, is its tendency to say that wisdom is a natural consequence of science. Over time this argument has served to reduce wisdom to the appliance of science in the aid of benign governance. Correspondingly, it has also served to devalue the qualities classically associated with wisdom, which as the above list illustrates had been largely defined in contrast to the values of science. For example, the two major modern ethical traditions, Kantianism and utilitarianism, demote the signature practical virtue of the wise, prudence: The prudent person may avoid doing harm but fails to perform optimally from the standpoint of a truly rational moral agent. Not surprisingly, both Kant and Bentham, in their rather different ways, took the structure of scientific reasoning as the template for the structure of moral reasoning.

It is not difficult to understand how this devaluation of wisdom occurred. Consider Francis Bacon's interpretation of Aristotle's conception of *phronesis*, or practical wisdom. This is the sort of art that, say, a successful politician possesses. Aristotle himself saw it as the product of long and attentive experience that one acquires from being a member of society and contributing to its various functions. For his part, Bacon focussed on the politician's ability to elicit the hidden potential in people, to enable them to be more than they might be otherwise by appealing to their underlying motives. This too requires considerable experience with people, but it involves viewing them in a certain way, more like objects that might be manipulated to one's advantage under certain conditions. Aristotle, of course, does not deny this aspect of politics. In fact, his works on rhetoric addressed them. However, much more than Bacon, Aristotle is concerned with the character of the politician and the appropriateness of his words to the situation. In contrast, for Bacon, the tasks of the sophist and the scientist converge in his version of practical wisdom: Politicians should learn to say and do what works.

An illuminating way to characterise the difference between Aristotle and Bacon on wisdom is in terms of the different senses of *immanence* and *transcendence* they foster. For Aristotle, the wise person appears immanent in the world in which he acts. The politician succeeds by being recognised as one of the people he would rule, a *primus inter pares*. His concerns are their concerns. This requires that he minimises aspects of his thought or being that might differentiate him from them. In that respect, he must transcend his own uniqueness. I put the point this way because Aristotle's sense of wisdom resonates with so-called Oriental Wisdom that stresses an embedded and contextualised understanding of oneself in the world more generally (e.g. Nisbett 2002). In contrast, the Baconian politician's sense of immanence involves his becoming one with his objective and using any means at his disposal to achieve it. He thus suspends whatever common feeling he might have with the people he would rule in order to view them in a way that enables him to get them where he wants to go. The relevant sense of transcendence here is that of a second-order observer capable of manipulating objects in a system. The corresponding sense of wisdom is that of an expert game player or sportsman.

The strong temptation to objectify subjects when acting from the Baconian standpoint is often condemned as an excess of the secular Enlightenment. Yet, as we have seen, such 'inhumanity' was routinely attributed to God in the supposedly more 'enchanted' world that practiced theodicy. Its guiding idea was that God's operations are constrained by our imperfect material natures, which means that the service of good may require the imposition of suffering. This perspective was recast in popular secular terms by Hegel as 'the cunning of reason', according to which the often ironic twists in the fate of individual human lives provide the means by which humanity as such comes to be full realised. This was the spirit in which Hegel endorsed Lord Bolingbroke's saying that history is philosophy teaching by examples. The retrospective vision provided by such lessons imparts its own brand of wisdom, but only to the spectators, never to the actors themselves. Specifically, it rationalises their plight and presumably puts their minds at ease, for no matter how badly things turn out, it has always been for a greater good.

Wisdom also has not fared well in the modern era because, as already suggested, it is biased in favour of the old and experienced. This attitude is clearly related to the modern suspicion, sometimes contempt, attached to tradition, which comes to be associated with mindless repetition, if not downright irrationality. Thus, to the modern mind, the prospect of losing or corrupting the accumulated knowledge of the past in succeeding generations is regarded as more blessing than curse. In the Enlightenment, Edward Gibbon and David Hume gave a decidedly positive spin to what had been previously regarded as the barbaric burning of the Library of Alexandria by the Caliph Omar in 642 AD

(Fuller and Gorman 1987). In this context, the old – understood both in terms of particular individuals and dominant social groups – are seen as blocking entire generations and classes of people who promise to bring fresh and different perspectives to the world. Starting with Plato, a minor theme in the history of political thought has been concerned with balancing the competing horizons of the young and the old in a fair system for selecting society's rulers. For example, Friedrich Hayek (1978) proposed that people vote once, in middle age, presumably the point when the natural liberality of youth has been tempered, but not entirely extinguished, by the lessons of age.

Thomas Kuhn (1970) made an interesting contribution to this general discussion of intergenerational epistemology, which may turn out to have been his most significant insight. If wisdom is an epistemic virtue associated with advancing years, the complementary virtue in the young is the ignorance of precedent that enables them to be effortlessly – some would say, witlessly – open-minded, which is to say, open to bold courses of thought and action. Kuhn deployed this point to argue that paradigm-shifts in science occur not because opponents of the same generation manage to reach agreement but because the younger generation of scientists supports the challenger to the orthodoxy. As Kuhn observes, these people are less personally invested in the old paradigm, not having directly experienced the benefits it provides in intellectual or professional terms. So they are more likely to take the old paradigm's long unsolved problems, combined with its relatively rigid authoritarian stance, as *prima facie* grounds for rejection. At the same, Kuhn realises that this open-mindedness can equally serve the orthodoxy – not least the incoming one – as the next generation of students are presented with an airbrushed 'Whig' account of their discipline's history that obscures the conceptual limitations and foregone alternatives that someone with a longer historical memory would naturally possess (Fuller 2003).

However, the various disappointments and disasters of self-identified 'modernist' projects over the past 250 years have periodically triggered sceptical reactions to modernity itself. The 'postmodern condition' is only the latest of a series of these reactions, which overall have increasingly turned for intellectual sustenance to the 'wisdom religions' of the East – Hinduism, Buddhism, Confucianism and Taoism (Clarke 1997). Before elaborating on the significance of this development, I want to dwell briefly on an additional wisdom-friendly factor of the postmodern condition – namely, its occurrence at a time when people's life expectancy has never been longer. Never before has there been sheer demographic pressure to promote the care and respect of the elderly. In service of this end, medical research has focussed considerable attention on arresting, if not reversing, Alzheimer's disease. But also there has been a revaluation of what in the past would have been seen as the spontaneous scepticism of the old towards youthful initiatives. The response of 'We've been here before' is now re-spun as a 'fast and reasonably reliable inductive judgement' (Goldberg 2005).

When the great world religions were formally classified in the 19th century, the difference between the West and the East was defined in terms of faiths based on *prophecy vs. wisdom* (Masuzawa 2005). The prophetic religions are the ones descended from the Old Testament – Judaism, Christianity, and Islam. They feature a single overpowering God in terms of whom humans are privileged by virtue of having been created in his 'image and likeness'. The prophetic religions are future-oriented. They regard secular life as an imperfect material condition that nevertheless provides the means by which a divinely planned order may be achieved. Various theological disciplines, including theodicy (on justice), soteriology (on salvation) and eschatology (on ultimate ends), were created to deal with the various issues raised by the underlying idea of reality as literally a 'work in progress'. In contrast, the wisdom religions do not accord their signature writings the same sacred status, largely because writing itself is not a privileged activity, since humans do not enjoy any special access to a supreme deity. Insofar as reality is subject to a creative force, it is one that rules humans in much the same sense as the rest of nature. In that case, wisdom comes from disciplining, if not entirely eliminating, any thoughts of overcoming this situation, let alone turning it to our own advantage: Instead of mastering the world, we become one with it.

All modern ideologies of progress owe at least a rhetorical debt to the prophetic religions. As I hinted above, a complementary debt is increasingly owed to the wisdom religions for the periodic turns against those ideologies. This became especially clear in the aftermath of the science-backed atrocities of the First World War. From popular works like Oswald Spengler's *Decline of the West*, wisdom emerged from cultures that felt their age (*Weltschmerz*) – and that it was time for the West to take a page from the East and feel its own age. A culture in its youth might legitimately risk its people and resources on dreams because the stakes are still relatively low. However, there comes a point when the appropriate response to failure is not to try harder but to relent and, as the Freudians say, 'cope' – that is, adapt to the most likely conditions and not desire or expect anything more. Wisdom is thus achieved as a kind of lived equilibrium with the environment. A subtle consequence of this value shift from the prophetic to the wisdom religions – also evident in Freud's work – is that the rejection of a transcendent 'father-like' deity came to be taken as a mark of the 'maturity' associated with wisdom. One does not become or replace the father, as in Kant's definition of Enlightenment as our release from 'nonage'. Rather, we come to realise that the father had never existed, and that the quest had been a waste of time and effort.

Over the past two centuries, the respective images of prophecy and wisdom as paradigms of religious knowledge have fluctuated. Originally the distinction was made clearly to the advantage of the prophetic religions, reflecting the tendency of the first wave of critical-historical scholarship of the Bible in the 18th century to group Judaism, Christianity and Islam together as species of 'monotheism', the religious precursor of the universalistic humanism that characterised Enlightenment thought. In this context, Islam was sometimes even portrayed as a purer form of monotheism than Christianity, given its more consistently spiritualised conception of the supreme deity, Allah. However, by the early 19th century, a literally 'Anti-Semitic' turn was taken by German philologists who became convinced that an early version of Sanskrit was the root of what came to be known as the 'Indo-European' languages. The Hebrew and Arabic of the Jews and Muslims belonged to a different linguistic-cum-ethnic heritage that potentially contaminated the line of descent that made its way from the Aryan regions of North India to Greece, Rome, Christendom and modern Europe. In this context, Plato's caste-like approach to politics and metaphysics started to look like the Buddha's refinements of Hinduism, both conforming to the contemplative mode of being characteristic of the wisdom religions. Correspondingly, Judaism and Islam as standard bearers of the prophetic religions started to appear barbaric and increasingly unfit for modern civil society. This is the image that evolved over the 19th century into what Edward Said (1978) notoriously dubbed 'Orientalism', an important intellectual source of recent forms of Anti-Semitism and Islamophobia.

In sum, then, the prophetic religions live ever in anticipation, the wisdom religions ever in adaptation. In terms of psychological paradigms, the former strives to complete a Gestalt, the latter to match a template. If prophetic religions appeal to youthful ambition, wisdom religions speak to the lessons of age. These differences suggest a further interesting difference between prophecy and wisdom over how to interpret free will: that we can get what we want *vs.* that we want what we can get. This distinction throws into sharp relief the least attractive feature of wisdom, one normally associated with ageing, namely, recognition that one's sphere of freedom has contracted. In other words, the 'wise' person operates with an increasingly path-dependent view of the world, such that whatever one does in a given situation, the result will fall within a predictably narrow range of outcomes. In youth we might shoot for the moon, but as adults we adapt to circumstances. In this respect, the difference between wisdom and cynicism is simply a matter of tact. Cynics ridicule those who do not yet share their wisdom, whereas the truly wise do not. In this particular comparison, I believe that the cynic's pre-emptive ridicule is the superior form of moral therapy, though others may wish to uphold the virtues of becoming wise 'the hard way'. But of course, I do not believe that either position really deserves our approval. Both are 'adaptive preferences', a phrase Elster (1983) imports from Leon Festinger's cognitive dissonance theory. It is better known as 'sour grapes', after the moral lesson of one of Aesop's fables. In the face of likely defeat or refutation (based on previous experience), the wise person simply re-specifies her values to aim at goals that are more achievable, thereby minimising her chances for disappointment.

In what follows, I focus exclusively on secular descendants of the prophetic vision, those who interpret our lack of existential fulfilment – Hegel's unhappy consciousness – as a spur to further action, and possibly even self-transcendence. In the modern period, the prophetic vision has morphed from the millenarian sect to the revolutionary party, generating four distinct sociodicies, or senses of redistributing human sentiment in aid of promoting a 'new world order':

(1) We might seek to establish a more durable version of the old order by strategies – from self-imposed Puritanism to society-wide eugenics – that prevent its likely sources of corruption.
(2) We might slow down the pace of progress in the advancement of humanity until the poor catch up through various taxation and incentive schemes in the name of social welfare.
(3) We might simply admit the defeat of the meliorist projects outlined in (2) and redefine the presence of the poor as 'overpopulation', an obstacle to progress that should be contained and ultimately eliminated within legally permissible constraints (aka 'letting die').
(4) We might seek a new currency of social value in terms of which everyone might trade in their old identities for new ones that place a smaller material demand on their recognition and thereby making it within more people's reach.

What happens if – or once – we reach Hegel's sense of the *end of history*, whereby thought and action are no longer artificially separated in the state of 'unhappy consciousness'. Given Hegel's own theological training, this is usefully seen as a secular resolution to the problem of Original Sin, whereby our reunion with God and the realisation of our species-defining ideals coincide with a just social order. From this perspective, heretical within Christianity, the sustaining belief in a heavenly afterlife that motivates our mundane struggles would turn out to have been itself a product of unhappy consciousness, a symptom of our distance from the unity of thought and action.

Life at the end of history was the main theme of the philosophical career of Alexandre Kojève, the European Economic Community bureaucrat who was France's answer to Georg Lukács in terms of retrofitting Hegel into a prophet of Marxism's early 20th century political ascendancy (Drury 1994). Here Kojève followed, albeit rather ironically, in the footsteps of Friedrich Nietzsche's own derisive speculations about Hegel's 'last man' in his 1872 prose poem *Thus Spake Zarathustra*, which launched the career of the *Übermensch*, the 'transhuman' being who would forever resist Hegel's 'final solution' to history. Nietzsche started by accepting Hegel's hypothesis that the end of history lay in a perfected Prussian welfare state – or if not, then a version of the emerging United States of America, with its substantial number of German settlers in a federal republic that appeared to strike a decent balance between personal freedom and political solidarity. But to Nietzsche, this supposed utopia amounted to an egalitarianism characteristic of what the social statistician Karl Pearson would soon call 'regression to the mean', whereby individual differences are flattened out, as everyone came to lead a similarly safe and contented existence. Thus, in Nietzsche's eyes, Hegel would have us all become cows who spend their days grazing. This reading of Hegel was given a remarkably euphemistic facelift in Fukuyama's (1992) best selling *The End of History and the Last Man*, in which the cows become consumers (cf. Fuller 2006b: chap. 9).

To be fair, Hegel's original idea was a secular version of Christian salvation, a literal 'heaven on earth' in which humans no longer had to struggle against the limits of their animal nature, presumably because everyone would enjoy a more than adequate standard of living. Implicit in this utopia was the idea that a perfect human life is one that could be lived with impunity – that is, without serious regard for the material consequences of what one did, said or thought. Only then would one be a genuinely 'free spirit'. Nevertheless, Nietzsche derided the boring character of such a world, implying that our humanity is intimately bound up with always having to struggle against our animal nature. Thus, for Nietzsche, reaching the end of history would result not in our self-realisation but our self-annihilation. We would bore ourselves to death, a fate that at least some critical media theorists believe that we have already reached (Postman 1985).

Nietzsche and Pearson, both influenced by Darwin, understood the dreaded homogeneity of Hegel's 'last man' in genetic as well as behavioural terms. Thus, at least in his Zarathustrian guise, Nietzsche deemed the *Übermensch* a worthy target of eugenically informed procreation. But writing a generation before the fundamental principles of heredity had been agreed, Nietzsche was unsurprisingly unclear about the details of his proposed solution to the prospect of perpetual mediocrity: Is the *Übermensch* a new and enhanced version of the human condition, as the English translation 'superman' suggests and today's biotech therapies promise? Or does it mark a return to some original moment of species excellence, an era of warriors and heroes that preceded our collective moral decline? In either case, while Nietzsche tried to portray the *Übermensch* as the ultimate value creator, it is hard to avoid the conclusion that such a being would leave a trail of violence and destruction in its wake. After all, along with 'the last man' who peacefully adapts to a world that allows the satisfaction of human desires, the *Übermensch*'s other great opponent is the Christian ascetic who in equally peaceful terms renounces his earthly existence in favour of some fabricated spiritual one. In contrast, the *Übermensch* is forever compelled to struggle for the fate of this world, perhaps even if that places him at odds with the humans in whose name he supposedly strives – as in the 1980s comic book series and recent film, *Watchmen*, which are predicated on a world that bans superheroes as part of an anxious Cold War strategy to maintain global peace.

If Nietzsche ironised Hegel's 'last man', Kojève ironised Nietzsche's *Übermensch*. Kojève developed many of his distinctive views in correspondence with the classical political theorist Leo Strauss shortly after the end of the Second World War. These are compiled with Strauss' essay *On Tyranny*, which was written in this period (Strauss 2000). Much to Strauss' consternation, Kojève portrays the historic resolution of Hegel's unhappy consciousness as a gradual blurring of the role of the philosopher and the 'tyrant' (understood abstractly as the ultimate source of power in society), as the latter learns ways of concretising – and thereby defining – the former's ideals. Kojève goes so far as to suggest that 'humanity' as a concept will come to have little meaning at the end of history, since the struggle of matter and spirit, so to speak, exemplified in the gulf between the tyrant and the philosopher, will have come to an end: 'Tyrants' will administer to 'subjects' whose desires they can meet, thereby quelling any need to revolt. In such a world, the residual emotions associated with the now resolved world-historic struggle will be domesticated in forms of 'gratuitous negativity', that is, random acts of violence that leave no lasting mark because they or memory of them can be always reversed (Drury 1994: chap. 4). This state-of-affairs should be seen as contributing to the trend, noted in the first section of this chapter, of the just society enabling a wrong to be righted by making it appear 'as if it never happened' – to quote the title of the landmark legal study in this area (Ripstein 2007).

A good sense of what Kojève had in mind may be got from watching Stanley Kubrick's film adaptation of the Anthony Burgess novel, *A Clockwork Orange*. As the protagonist Alex repeatedly exemplifies, such a world is without the experience of regret – or, for that matter, wisdom, whose powers of hindsight depend crucially on leveraging our incapacity to change the past into an adaptive strategy for the future (Tierney 2009). Kojève sometimes spoke about this as the 'reanimalisation of man', which suggests a reabsorption of humanity into nature, a point originally symbolised in Burgess' use of Malay slang, where 'orange' implies the equation of humans and orangutans. It is worth stressing that within the history of Marxism, Kojève's general line of thought is itself not peculiar – only the spin he gives it. Seeing with the eye of a bureaucrat rather than a revolutionary, he nevertheless follows Leon Trotsky, who also valorised the self-expressive power of destruction (albeit Trotsky had more in mind modern art than reparable vandalism) and regarded wisdom as ultimately a mark of the weak, or at least incomplete, state of the human condition (Molnar 1972: 100).

5 The dawn of suffering smart: Recycling evil in the name of good

There are two ways to think about the role of suffering in the narrative of humanity, each of which provides an intuitive benchmark for progress. On the one hand, one might aim to minimise suffering as an end in itself, which in turn requires eliminating the conditions that bring about suffering in the first place. In this category would fall Epicurean and other therapeutic philosophies that aim to relieve our distress about things over which can exert no substantial control. Progress in life then amounts to coping better by keeping one's desires in line with reasonable expectations. Such a trajectory also leads one to think in terms of an 'adequately lived life', after which suicide appears as a dignified means to prevent an increase in one's own and the world's misery. In recent times Peter Singer is probably the philosopher who has most publicly and comprehensively embraced this perspective.

On the other hand, one might think in terms of learning to tolerate greater suffering as a short-to-medium term cost for producing more sustainable long-term benefits, as in Nietzsche's motto, 'What does not kill me makes me stronger', or for that matter, the transhumanist imperative to re-engineer the human body to enable us to live longer so as to work and play harder – as previously discussed in Chapter 3. This way of thinking is more in the spirit of classical theodicy, whereby progress is ultimately about the sublimation rather than the outright elimination of suffering. In Christianised cultures, especially after the Protestant Reformation, it has meant living creatively with Original Sin, converting private vices into public benefits, to recall the first formulation of the invisible hand as an account of commercial life: To achieve one's own ends, which may be quite exclusive, one must first do something that benefits others, which may turn out to be quite inclusive. The science of economics has followed through on this perspective most thoroughly to project the image of a forever dynamic and expanding society. However, the ultimate source of this vision has been the figure of Jesus who, quite unlike Socrates, actively cultivated suffering – even took on the suffering of others – in order to fulfil his divine mandate. The difference, of course, is that the end pursued by Jesus was maximally inclusive (i.e. salvation), while the means he was required to pursue was exclusive to himself (i.e. death on the cross).

To appreciate the role that the 'long term' plays in theodicy's moral calculus, it is worth recalling that, despite their standing amongst the very noblest of humans, Socrates and Jesus was each deemed a failure in the period immediately following his death. In both cases, his most distinguished follower – Plato and St Paul – avoided the cities that were supposed to have been revolutionised by his master's words, whilst those cities – Athens and Jerusalem – fell into disarray and decline. In perhaps the most profound secular treatment of theodicy in recent times, Barrington Moore (1970) similarly noted that in the first 50 years after the 1789 French Revolution, it was easy for liberals to share Alexis de Tocqueville's negative verdict on its efficacy. However, especially after 1848, as it became clear that traditional European elites were unlikely to yield to democratic reforms without at least the threat of force, the French Revolution acquired a talismanic quality that politicians and theorists of the Left continued to promote in the 20th century – arguably even today. The point is that the undeniable suffering associated with the original event comes to be seen as a necessary part of an optimal solution to a perennial problem in the human condition. The passage of time allows the morally discerning to appreciate the world-historic constraints on the possibilities for constructive action, so that what at first seemed to be an unnecessarily violent episode turns out to have been a necessary wake up call.

Lest this appraisal appear too 'panglossian' to be morally credible, consider that the person regularly regarded as the greatest American, Abraham Lincoln, presided over a civil war that was by far the worst conflict in that nation's history. The sheer number of military casualties alone outstripped those of the second bloodiest war – the Second World War – by a 3:2 ratio. Considering that the two wars lasted roughly the same amount of time (four years) yet the nation's population had grown fourfold in the 80 year interval, the Civil War arguably had six times the impact as the Second World War, which is normally regarded as having left lasting scars on American society. Nevertheless, Lincoln's emancipation of the slaves, reassertion and expansion of civil rights (at least in principle) and prevention of any subsequent civil wars have stood the test of time, especially in a period that witnessed America's shift from a hemispheric to a global power. To be sure, it would be foolish to credit Lincoln with having intended all of these results in quite the way it turned out but it is difficult to see in retrospect how the US could have acted as decisively as it has on the world stage, were it not sufficiently unified to resist the sort of divisive foreign influences that would otherwise plague a nation of multiple immigrant populations and long-standing urban-rural divides. All things considered, then, the US Civil War has come to be seen as a 'good thing', a golden opportunity to fortify its civic republican immune system.

Something similar – but on an international scale – has been argued about the ultimate moral significance of US aerial bombing of German and Japanese cities to expedite an end to the Second World War. In effect, the deployment of the 'two wrongs make a right' principle appears to have immunised the planet from the onset of any more world wars – at least in the style of the two that punctuated the 20th century. The United Nations, notwithstanding its periodic deficiencies of will, remains the great symbol of that legacy. The tricky part of putting matters this way is how, if at all, to rationalise the first wrong, the Original Sin, so to speak. After all, the 'two wrongs make a right' principle has the basic structure of a Hegelian dialectic, in which two inadequate entities cancel out each other's differences to produce a (more) complete entity. Implied here is that there was some latent good represented by, say, the

Confederate States of America or Nazi Germany that deserves to be preserved in sublimated form in some future ideal or at least improved version of the human condition. In the case of the Confederate States, whose secession from the US triggered the Civil War, historians have long accepted the Jeffersonian anti-federalist, pro-states' rights and popular sovereignty animus that has always been an important current in America's self-understanding, which continues to this day in the libertarian wing of the Republican Party, as recently revived by the Tea Party Movement. Albeit often in defence of odious policies, this strand of the American political tradition has consistently upheld the legitimacy of social action taken by self-organising individuals. Can such a story be told even for Nazi Germany, in spite of the enormity of the suffering it caused?

If we wish to continue including the Nazis as part of the history of humanity – as opposed to the history of nature – then the answer must be yes, however difficult at first glance that may seem. Put bluntly, we must envisage the prospect of a transformation in the normative image of Nazi Germany comparable to what Barrington Moore described for the French Revolution. This is not easy. The makeover Moore observed occurred over a couple of generations, and in that same amount of time there have been only the barest hints of Nazi rehabilitation. But hints there are, helped along by the death of those with first-hand experience of Nazism. To be sure, some areas of Nazi science that did not figure prominently in the Second World War – such as space travel, ecology and cancer research – were easily, if somewhat surreptitiously, assimilated by the Allies. But even in the case of the Nazi science of 'racial hygiene', there is a dawning realisation that 'eugenics' and 'genetic modification' more generally have been always integral to progressive normative agendas. In that case Nazi science policies are perhaps best seen as opportunistically extreme versions of tendencies long present and accepted by the intellectual vanguard of Western culture. Lest this speculation seem, once again, too panglossian, it is worth noting that Nazi Germany promoted itself in just this way – with considerable success in the international media – before the presentation of evidence for the Holocaust. If one is inclined to think, as I am, that the Holocaust was produced by the exigencies of war rather than intrinsic to the Nazi agenda, then a key to recovering the 'good' in Nazism might be to rewind history to the 1920s and 1930s when the movement appeared to offer the promise of a progressive future. Back then the Holocaust did not appear to be an inevitable outcome of Nazism, which in turn enabled observers to see Nazis as fruitfully extending existing scientific agendas (cf. Fuller 2006b: chap. 14).

A fictional genealogy of theodicy would extend from Voltaire's satirisation of Leibniz as Dr Pangloss to Stanley Kubrick's Dr Strangelove (Fuller 2010: chap. 1). However, the period after the Second World War has thrown up some real-life exemplars of these figures, all of whom conform to J. Robert Oppenheimer's likening of his own role in producing the first successful atomic bomb to the Hindu god Shiva, the destroyer and creator of worlds. Moreover, the allusion to Shiva echoes Joseph Schumpeter's characterisation of the entrepreneur as the 'creative destroyer' of markets. With those associations in mind, consider what might be called the *moral entrepreneurship* of the likes of Robert McNamara, Jeffrey Sachs and George Soros. Each in his inimitable way has managed to leverage evil into good, arguably because his commission of evil demonstrated a level of competence in managing global affairs that, under the right circumstances, could be turned to good ends. Thus, McNamara went from orchestrating US involvement in the Vietnam War to boosting the World Bank's development remit, Sachs from introducing harsh neo-liberal reforms to former Soviet republics to the United Nations' chief envoy for global poverty reduction and Soros from arch currency speculator to eloquent advocate of global financial regulation. The take-home lesson here may be that a sense of urgency is needed for people of good knowledge to become people of good will (as opposed to people of bad will or, in the case of pure academics, no will whatsoever). The resulting rehabilitation – if not resurrection – would be a very concrete way of bridging the gap between theory and practice resulting from Hegel's unhappy consciousness.

The 21st century revival of theodicy will involve the institutionalisation of moral entrepreneurship, as an increasing portion of the human condition becomes reversible, through either literal regeneration of body parts or the provision of functionally equivalent states of being. The former is most evident in the scientific attempts to prolong productive human lifespan by reversing the ageing process and supplying prosthetic enhancement of our normal capacities. (We have considered these motives and prospects for these projects in Chapter 3.) The latter is manifested in a wide variety of typically money-based schemes that aim to provide a common standard of measurement for states of being that otherwise would be seen as incommensurable. The payout of damages in lieu of incarceration by a defendant found guilty already opened the door to this arrangement (Ripstein 2007). But we may add proactive schemes to create markets in, say, carbon emissions, whereby heavy polluters would be obliged to make deals with low polluters in order to ensure the world's carbon sustainability. The idea may be even extended to indices of 'cognitive sustainability', whereby the persistent intellectual asymmetries, including 'brain drains', between nation-states could be redressed by a global tax on the education-rich countries hypothecated to the cultivation of intellectual capital in education-poor countries. In effect, 'affirmative action' and 'positive discrimination' would enjoy trans-national political status.

Moral entrepreneurship normalises evil by recycling it into good. To appreciate how this already happens at a global level, consider investors who make a killing in the currency, precious metals or energy markets whenever a natural or moral catastrophe strikes. From the God's eye view of financial capitalism, these beneficiaries of others' misfortunes are no mere predators but spontaneous wealth re-distributors. They personally benefit in the short term but in the long term their taxes and further investments that follow in the wake of their seemingly ill-got gains serve to re-capitalise the economy. Indeed, the very people who had been harmed by the catastrophe may stand to gain – if they can hold on long enough! This narrative is certainly familiar from the long-term structural advantage of war for defeated societies that manage to re-build themselves by incorporating the know-how of their former enemies, who are typically their largest investors. Against this back-drop, the forms of global governance suggested by, say, carbon trading schemes, levies on currency transactions (the so-called Tobin Tax) or my own trans-national epistemic tax are best seen as attempts to render the outcomes of moral entrepreneurship somewhat more accountable, if still not entirely predictable. In particular, our sense of justice is tied to redistributions being conducted in a *timely, targeted* and *proportional* fashion. What the italicised terms mean in the 21st century will demand from social theorists unprecedented levels of realism, imagination and will.

It is worth underscoring this last point. After all, if carbon emissions rise at their current rate globally, then a massive redistribution of the world's resources may well occur – one involving the death and destruction of many human societies that were not themselves major polluters yet still unable to protect themselves from major pollution. In this scenario, the largest polluters would continue to survive and even thrive, albeit with structural adjustments to their socio-economic orders. The injustice here is more keenly felt if we think of each human being as a member of a species with whom one shares certain essential qualities than as part of a population within which one would expect a finite amount of variation amongst individuals. In the former case, the sense of human equality, and hence empathy, is much stronger. In any case, the difference in perspective encapsulates how the ontology of life looks before and after Darwin (Mayr 1994). The Darwinian view today does not seem quite as harsh as it did a century ago, when it travelled under the rubric of 'Social Darwinism'. That shift in sentiment has had less to do with acceptance of evolutionary biology than with a curious hybridisation of the old species view and the new population view of humanity, courtesy of global information and communication technologies that instil a sense in even the poorest and most vulnerable person that she too is a consumer with choice over her fate. But that choice may involve transferring one's material and spiritual resources – and ultimately one's life – to some other entity, be it a state with imperial ambitions or an avatar in cyberspace. In short, 'suffering smart' may be less about extending your current mode of existence than exchanging it for one with a greater chance of achieving your aims more effectively. In that respect, the moral horizons of Humanity 2.0 are about defining what is in need of continual resurrection.

Bibliography

Adleman, L.M. (1994) 'Molecular computation of solutions to combinatorial problems'. *Science*, 266: 1021–4.

Agamben, G. (1998) *Homo Sacer: Sovereign Power and Bare Life*. Palo Alto: Stanford University Press.

Agamben, G. (2005) *The State of Exception*. Chicago: University of Chicago Press.

Ainslie, G. (1992) *Picoeconomics: The Strategic Interaction of Successive Motivational States within the Person*. Cambridge UK: Cambridge University Press.

Ainslie, G. (2001) *Breakdown of Will*. Cambridge UK: Cambridge University Press.

Amos, M. (2006) *Genesis Machines: The New Science of Biocomputing*. London: Atlantic Books.

Anderson, B. (1983) *The Imagined Community*. London: Verso.

Anderson, W.T. (2003) 'Argumentation, symbiosis, transcendence: Technology and the future(s) of human identity'. *Futures*, 35: 535–46.

Armstrong, K. (2009) *The Case for God: What Religion Really Means*. London: Bodley Head.

Armstrong, R. (2009) 'Protocells and plectic systems architecture'. *Sophia*, 3 (June): 17. http://www.ucl.ac.uk/~ucbpeal/sophia/issue3-web.pdf

Arthur, W.B. (1994) *Increasing Returns and Path Dependence in the Economy*. Ann Arbor MI: University of Michigan Press.

Atkinson, Q.D. and R.D. Gray (2005) 'Curious parallels and curious connections – phylogenetic thinking in biology and historical linguistics'. *Systematic Biology*, 54(4): 513–26.

Ayala, F. (2007) 'Darwin's greatest discovery: Design without a designer'. *Proceedings of the National Academy of Sciences*, 104 (May): 8567–73.

Baber, Z. (ed.) (2009) 'Review Symposium on Steve Fuller's *The New Sociological Imagination*'. *History of the Human Sciences*, 22: 110–45.

Baillie, H. and T. Casey (eds) (2004) *Is Human Nature Obsolete?: Genetics, Bioengineering, and the Future of the Human Condition*. Cambridge MA: MIT Press.

Bainbridge, W.S. (2005) 'The transhuman heresy'. *Journal of Evolution and Technology*, 14(2): 1–10. Available from: http://jetpress.org/volume14/bainbridge.html

Ball, P. (2003) 'Nanotechnology in the firing line'. http://www.nanotechweb.org/articles/society/2/12/1/1, 23 December.

Ball, P. (2004) *Critical Mass: How One Thing Leads to Another*. New York: Farrar Straus Giroux.

Barben, D., E. Fisher, C. Selin, D. Guston (2008) 'Anticipatory governance of nanotechnology: Foresight, engagement and integration', in E. Hackett, O. Amsterdamska, M. Lynch, J. Wacjman (eds) *Handbook of Science and Technology Studies*, pp. 979–1000. Cambridge MA: MIT Press.

Barnatt, C. (2010) *A Brief Guide to Cloud Computing*. London: Constable and Robinson.

Basalla, G. (2006) *Civilized Life in the Universe: Scientists on Intelligent Extraterrestrials*. Oxford: Oxford University Press.

Behe, M. (1996) *Darwin's Black Box*. New York: Simon and Schuster.

Bellamy, R. (1992) *Liberalism and Modern Society*. Cambridge: Polity.

Benyus, J. (1997) *Biomimicry*. New York: William Morrow.

Berger, P. and T. Luckmann (1968) *The Social Construction of Reality*. Garden City NY: Doubleday.

Berman, S. (2006) *The Primacy of Politics: Social Democracy and the Making of Europe's Twentieth Century*. Cambridge UK: Cambridge University Press.

Bernal, M. (1987) *Black Athena*. New Brunswick NJ: Rutgers University Press.

Bhaskar, R. (1975) *A Realist Theory of Science*. Brighton UK: Harvester.

Bhaskar, R. (1986) *Scientific Realism and Human Emancipation*. London: Verso.

Bibel, W. (ed.) (2004) 'Converging technologies and the natural, social and cultural world', Special Interest Group Report for the European Commission via an Expert Group on 'Foresighting the New Technology Wave' (26 July), ftp://ftp.cordis.europa.eu/pub/foresight/docs/ntw_sig4_en.pdf

Bijker, W. and J. Law (eds) (1993) *Shaping Technology/Building Society*. Cambridge MA: MIT Press.

Böhm-Bawerk, E.v. (1959[1884]) *Capital and Interest: History and Critique of Interest Theories*. South Holland, IL: Libertarian Press.

Bowler, P. (1989) *The Mendelian Revolution*. Baltimore: Johns Hopkins University Press.

Brague, R. (2007) *The Law of God: The Philosophical History of an Idea*. Chicago: University of Chicago Press.

Brams, S. (2002) *Biblical Games: Game Theory and the Hebrew Bible*. Cambridge MA: MIT Press.

Brantley, R. (1984) *Locke, Wesley and the Method of English Romanticism*. Gainesville FL: University of Florida Press.

Brattain, M. (2007) 'Race, racism, and antiracism: UNESCO and the politics of presenting science to the postwar public'. *American Historical Review*, 112(5): 1386–413.

Brenner, R. (1987) *Rivalry: In Business, Science, Among Nations*. Cambridge: Cambridge University Press.

Brett-Crowther, M.R. (1981) 'Teilhard in retrospect'. *Futures*, 13(6): 517–20.

Broberg, G. and N. Roll-Hansen (eds) (1997) *Eugenics and the Welfare State: Sterilization Policy in Denmark, Sweden, Norway, and Finland*. Lansing MI: Michigan State University Press.

Brown, D.E. (1988) *Hierarchy, History and Human Nature: The Social Origins of Historical Consciousness*. Tucson: University of Arizona Press.

Bud, R. (1993) *The Uses of Life: A History of Biotechnology*. Cambridge UK: Cambridge University Press.

Burtt, E.A. (1925) *The Metaphysical Foundations of Modern Physical Science*. London: Kegan Paul.

Butler, J. (1990) *Gender Trouble*. London: Routledge.

Butterfield, H. (1949) *Christianity and History*. London: G. Bell and Sons.

Camcastle, C. (2005) *The More Moderate Side of Joseph De Maistre: Views on Political Liberty and Political Economy*. Montreal: McGill University Press.

Carroll, S.B. (2005) *Endless Forms Most Beautiful: The New Science of Evo-Devo*. New York: Norton.

Carroll, S.B. (2006) *The Making of the Fittest: DNA and the Ultimate Forensic Record of Evolution*. New York: W.W. Norton.

Cavalli-Sforza, L. (2000) *Genes, Peoples, and Languages*. New York: Farrar, Straus and Giroux.

Chang, K. (2003) 'Smaller computer chips built using DNA as template'. *New York Times*, 21 November.

Chatterjee, A. (2007) 'Cosmetic neurology and cosmetic surgery: Parallels, predictions and challenges'. *Cambridge Quarterly of Healthcare Ethics*, 16: 129–37.

Cheng, D., M. Claessens, T. Gascoigne, J. Metcalfe, B. Schiele, S. Shi (eds) (2008) *Communicating Science in Social Contexts: New Models, New Practices*. Berlin: Springer Science and Business Media.

Chomsky, N. (1966) *Cartesian Linguistics: A Chapter in the History of Rationalist Thought*. New York: Harper and Row.

Clark, A. (2008) *Supersizing the Mind*. Cambridge MA: MIT Press.

Clarke, J.J. (1997) *Oriental Enlightenment*. London: Routledge.

Cohen, I.B. (1995) *Science and the Founding Fathers*. New York: W.W. Norton.

Colley, L. (1992) *Britons: Forging the Nation, 1707–1837*. New Haven CT: Yale University Press.

Collini, S. (1979) *Liberalism and Sociology: L.T. Hobhouse and the Political Argument in England 1880–1914*. Cambridge UK: Cambridge University Press.

Collini, S. (2006) *Absent Minds: Intellectuals in Britain*. Oxford: Oxford University Press.

Collini, S., D. Winch, J.W. Burrow (1983) *That Noble Science of Politics: A Study in Nineteenth Century Intellectual History*. Cambridge UK: Cambridge University Press.

Collins, R. (1979) *The Credential Society*. New York: Academic Press.

Collins, R. (1998) *The Sociology of Philosophies: A Global Theory of Intellectual Change*. Cambridge MA: Harvard University Press.

Corbey, R. (2005) *The Metaphysics of Apes*. Cambridge UK: Cambridge University Press.

Coupland, P. (2000) 'H.G. Wells' "Liberal Fascism"'. *Journal of Contemporary History*, 35: 541–55.

Cox, B. (1997) 'HotWired BrainTennis Debate', November, http://virtualschool.edu/cox/pub/97WiredBrainTennis/

Cox, J.R. and C.A. Willard (eds) (1982) *Advances in Argumentation Theory and Research*. Tuscaloosa AL: University of Alabama Press.

Cremaschi, S. and M. Dascal (1996) 'Malthus and Ricardo on economic methodology'. *History of Political Economy*, 28(3): 475–511.

Dahrendorf, R. (1995) *LSE: A History of the London School of Economics and Political Science, 1895–1995*. Oxford: Oxford University Press.

Darwin, C. (1859) *On the Origin of Species by Means of Natural Selection*. London: John Murray.

Darwin, C. (1871) *The Descent of Man, and Selection in Relation to Sex*. London: John Murray.

Davies, P. (2010) *The Eerie Silence: Are We Alone in the Universe?* London: Allen Lane.

Davis, E. (1998) *TechGnosis: Myth, Magic and Mysticism in the Age of Information*. Harmony Books: San Francisco.

Dawkins, R. (1976) *The Selfish Gene*. Oxford: Oxford University Press.

Dawkins, R. (1982) *The Extended Phenotype*. Oxford: Oxford University Press.

Dawkins, R. (1999) 'Foreword', in J. Burley (ed.) *The Genetic Revolution and Human Rights*. Oxford: Oxford University Press.

Dawkins, R. (2006) *The God Delusion*. New York: Houghton Mifflin.
Dear, P. (2006) *The Intelligibility of Nature*. Chicago: University of Chicago Press.
DeGrey, A. (2007) *Ending Aging: The Rejuvenation Breakthroughs that Could Reverse Human Aging in Our Lifetime*. New York: St Martin's Press.
Deleuze, G. (1994) *Difference and Repetition* (Orig. 1968). New York: Columbia University Press.
Deleuze, G. and F. Guattari (1987) *A Thousand Plateaus* (Orig. 1980). Minneapolis: University of Minnesota Press.
Dembski, W. (1998) *The Design Inference*. Cambridge UK: Cambridge University Press.
Dennett, D. (1995) *Darwin's Dangerous Idea*. London: Allen Lane.
Desmond, A. and J. Moore (2009) *Darwin's Sacred Cause: Race, Slavery and the Quest for Human Origins*. London: Allen Lane.
Deutsch, K., A. Markovits, J. Platt (eds) (1986) *Advances in the Social Sciences, 1900–1980*. Lanham MD: University Press of America.
Dickens, P. (2000) *Social Darwinism: Linking Evolutionary Thought to Social Theory*. Milton Keynes UK: Open University Press.
Dobzhansky, T. (1937) *Genetics and the Origin of Species*. New York: Columbia University Press.
Dobzhansky, T. (1967) *The Biology of Ultimate Concern*. New York: New American Library.
Douglas, K. (2008) 'Six "uniquely" human traits now found in animals'. *New Scientist*, 22 May.
Drexler, E. (1986) *Engines of Creation: The Coming Era of Nanotechnology*. Garden City NY: Doubleday.
Drury, S. (1994) *Alexandre Kojève: The Roots of Postmodern Politics*. London: Palgrave Macmillan.
Duhem, P. (1954) *The Aim and Structure of Physical Theory* (Orig. 1914). Princeton: Princeton University Press.
Dupré, J. (1993) *The Disorder of Things*. Cambridge MA: Harvard University Press.
Dupuy, J-P. (2004) 'Complexity and uncertainty: A prudential approach to nanotechnology'. A contribution to 'Foresighting the new technology wave'. High Level Expert Group, European Commission, Brussels. http://portal.unesco.org/ci/en/files/20003/11272944951Dupuy2.pdf/Dupuy2.pdf
Durkheim, E. (2001[1912]) Th*e Elementary Forms of the Religious Life*. Oxford: Oxford University Press.
Dyson, F. (2007) 'Our biotech future'. *The New York Review of Books*, vol. 54, no. 12 (19 July).
Elster, J. (1983) *Sour Grapes: Studies in the Subversion of Rationality*. Cambridge UK: Cambridge University Press.
Fara, P. (2002) *Newton: The Making of Genius*. New York: Columbia University Press.
Feyerabend, P. (1975) *Against Method*. London: New Left Books.
Feyerabend, P. (1999) *The Conquest of Abundance*. Chicago: University of Chicago Press.
Fleischacker, S. (2004) *A Short History of Distributive Justice*. Cambridge MA: Harvard University Press.
Fodor, J. (1981) *Representations*. Cambridge MA: MIT Press.

Foucault, M. (1970) *The Order of Things: An Archaeology of the Human Sciences*. New York: Pantheon Books.

Franklin, A. (1999) *Animals and Modern Cultures: A Sociology of Human-Animal Relations in Modernity*. London: Sage.

Fukuyama, F. (1992) *The End of History and the Last Man*. New York: Free Press.

Fukuyama, F. (2002) *Our Posthuman Future: Consequences of the Biotechnology Revolution*. New York: Farrar Straus Giroux.

Fuller, S. (1988) *Social Epistemology*. Bloomington IN: Indiana University Press.

Fuller, S. (1993) *Philosophy of Science and Its Discontents*. 2nd edn (Orig. 1989). New York: Guilford Press.

Fuller, S. (1997) *Science*. Milton Keynes UK: Open University Press.

Fuller, S. (1998a) 'A social epistemology of the structure-agency craze: From content to context', in A. Sica (ed.) *What is Social Theory?: The Philosophical Debates*, pp. 92–117. Oxford: Blackwell.

Fuller, S. (1998b) 'Divining the future of social theory: From theology to rhetoric via social epistemology'. *European Journal of Social Theory*, 1: 107–26.

Fuller, S. (2000a) *The Governance of Science: Ideology and the Future of the Open Society*. Milton Keynes UK: Open University Press.

Fuller, S. (2000b) *Thomas Kuhn: A Philosophical History for Our Time*. Chicago: University of Chicago Press.

Fuller, S. (2002a) *Knowledge Management Foundations*. Woburn MA: Butterworth-Heinemann.

Fuller, S. (2002b) 'Making up the past: A response to Sharrock and Leudar'. *History of the Human Sciences*, 15(4): 115–23.

Fuller, S. (2003) *Kuhn vs Popper: The Struggle for the Soul of Science*. Cambridge UK: Icon.

Fuller, S. (2006a) *The Philosophy of Science and Technology Studies*. London: Routledge.

Fuller, S. (2006b) *The New Sociological Imagination*. London: Sage.

Fuller, S. (2007a) *New Frontiers in Science and Technology Studies*. Cambridge UK: Polity.

Fuller, S. (2007b) *Science vs. Religion? Intelligent Design and the Problem of Evolution*. Cambridge UK: Polity.

Fuller, S. (2007c) 'Learning from error: An autopsy of Bernalism'. *Science as Culture*, 16: 463–6.

Fuller, S. (2008a) *Dissent over Descent: Intelligent Design's Challenge to Darwinism*. Cambridge UK: Icon.

Fuller, S. (2008b) 'The normative turn: Counterfactuals and a philosophical historiography of science'. *Isis*, 99: 576–84.

Fuller, S. (2009a) *The Sociology of Intellectual Life: The Life of the Mind in and Around the Academy*. London: Sage.

Fuller, S. (2009b) 'Science studies goes public: Report on an ongoing performance'. *Spontaneous Generations*, 2(1): 11–21.

Fuller, S. (2010) *Science: The Art of Living*. Durham UK: Acumen.

Fuller, S. and D. Gorman (1987) 'Burning libraries and the problem of historical consciousness'. *Annals of Scholarship*, 4(3): 105–22.

Fuller, S. and J. Collier (2004) *Philosophy, Rhetoric and the End of Knowledge*, 2nd edn (Orig. 1993, by Fuller). Mahwah NJ: Lawrence Erlbaum Associates.

Funkenstein, A. (1986) *Theology and the Scientific Imagination*. Princeton: Princeton University Press.
Galbraith, J.K. (1958) *The Affluent Society*. New York: Houghton Mifflin.
Gale, R. (ed.) (1967) *The Philosophy of Time*. Garden City NY: Doubleday.
Galison, P. (1999) *Image and Logic*. Chicago: University of Chicago Press.
Galison, P. and D. Stump (eds) (1996) *The Disunity of Science: Boundaries, Contexts and Power*. Palo Alto CA: Stanford University Press.
Gee, H. (2000) *In Search of Deep Time*. London: HarperCollins.
Geras, N. (1998) *The Contract of Mutual Indifference*. London: Verso.
Gibbons, M., C. Limoges, H. Nowotny, S. Schwartzman, P. Scott and M. Trow (1994) *The New Production of Knowledge*. London: Sage.
Giddens, A. (1976) *New Rules of the Sociological Method*. London: Macmillan.
Giddens, A. (1984) *The Constitution of Society*. Cambridge UK: Polity.
Gilbert, W. (1991) 'Towards a paradigm shift in biology'. *Nature*, 10 January, 349, 99.
Gilder, G. (1989) *Microcosm: The Quantum Revolution in Science and Technology*. New York: Simon and Schuster.
Glover, J. (1984) *What Sort of People Should There Be?* Harmondsworth UK: Penguin.
Goldberg, E. (2005) *The Wisdom Paradox: How Your Mind Can Grow Stronger as Your Brain Grows Older*. New York: Free Press.
Goldberg, J. (2007) *Liberal Fascism*. Garden City NY: Doubleday.
Goldman, E. (1952) *Rendezvous with Destiny: A History of Modern American Reform*. New York: Alfred Knopf.
Goodman, N. (1954) *Fact, Fiction, and Forecast*. Cambridge MA: Harvard University Press.
Gould, S.J. (1988) *Wonderful Life*. New York: Norton.
Gould, S.J. (1999) *Rocks of Ages*. New York: Random House.
Gray, J. (2002) *Straw Dogs: Thoughts on Humans and Other Animals*. London: Granta Books.
Gray, J. (2007) *Black Mass: Apocalyptic Religion and the Death of Utopia*. London: Allen Lane.
Greenberg, D.D. (2007) *Science for Sale: The Perils, Rewards, and Delusions of Campus Capitalism*. Chicago, IL: University of Chicago Press.
Grenfell, M. (2005) *Pierre Bourdieu: Agent Provocateur*. London: Macmillan.
Habermas, J. (1981) *A Theory of Communicative Action*. Boston: Beacon Press.
Habermas, J. (2002) *The Future of Human Nature*. Cambridge UK: Polity.
Hacking, I. (1990) *The Taming of Chance*. Cambridge UK: Cambridge University Press.
Hacking, I. (1998) *The Social Construction of What?* Cambridge MA: Harvard University Press.
Haraway, D. (1991) *Simians, Cyborgs, Women*. London: Free Association Books.
Haraway, D. (2003) *Companion Species Manifesto*. Chicago: Prickly Paradigm Press.
Harman, O. (ed.) (2008) *Rebels, Mavericks and Heretics in Biology*. New Haven: Yale University Press.
Harmon, A. (2006) 'DNA hunters hit snag: Tribes don't trust them'. *The New York Times*, December 10.
Harris, J. (2007) *Enhancing Evolution: The Ethical Case for Making Better People*. Princeton: Princeton University Press.

Harris, M. (1968) *The Rise of Anthropological Theory*. New York, NY, USA: Crowell.
Harrison, P. (1998) *The Bible, Protestantism and the Rise of Natural Science*. Cambridge UK: Cambridge University Press.
Hauser, M., N. Chomsky, W. Fitch (2002) 'The faculty of language: What is it, who has it, and how did it evolve?'. *Science*, 298: 1569–79.
Hayek, F. (1952) *The Counter-Revolution in Science*. Chicago: University of Chicago Press.
Hayek, F. (1978) *New Studies in Philosophy, Politics and Economics*. London: Routledge & Kegan Paul.
Hecht, J.M. (2003) *The End of the Soul*. New York: Columbia University Press.
Heelan, P. (1983) *Space-Perception and the Philosophy of Science*. Berkeley CA: University of California Press.
Heims, S. (1991) *Constructing a Social Science for Postwar America: The Cybernetics Group, 1946–1953*. Cambridge MA: MIT Press.
Helmreich, S. (1998) *Silicon Second Nature: Culturing Artificial Life in a Digital World*. Berkeley: University of California Press.
Helpman, E. and M. Trajtenberg (1994) *A Time to Sow and a Time to Reap: Growth Based on General Purpose Technologies*. NBER working paper no. 4854. Cambridge MA: National Bureau of Economic Research.
Hesse, M.B. (1965) *Models and Analogies in Science*. South Bend IN: Notre Dame University Press.
Hirsch, F. (1976) *The Social Limits to Growth*. London: Routledge and Kegan Paul.
Hirschman, A. (1991) *The Rhetoric of Reaction*. Cambridge MA: Harvard University Press.
Hirst, P. (1975) *Durkheim, Bernard and Epistemology*. London: Routledge & Kegan Paul.
Hofstadter, R. (1944) *Social Darwinism in American Thought*. Philadelphia: University of Pennsylvania Press.
Hofstadter, R. (1965) *The Paranoid Style in American Politics*. New York: Alfred Knopf.
Hogben, L. (ed.) (1938) *Political Arithmetic*. London: Routledge & Kegan Paul.
Holliman, R., E. Whitelegg, E. Scanlon, S. Smidt, J. Thomas (eds) (2009) *Investigating Science Communication in the Information Age: Implications for Public Engagement and Popular Media*. Oxford: Oxford University Press and Open University Press.
Horgan, J. (1996) *The End of Science*. Reading MA: Addison-Wesley.
Hull, D. (1989) *The Metaphysics of Evolution*. Albany NY: SUNY Press.
Hunter, P. (2008) 'What genes remember'. *Prospect*, 146 (May).
Ingold, T. (1994) 'Humanity and animality', in T. Ingold (ed.) *Companion Encyclopedia of Anthropology*. Chap. 2. London: Routledge.
Israel, J. (2001) *Radical Enlightenment*. Oxford: Oxford University Press.
Jacoby, R. (1987) *The Last Intellectuals*. New York: Basic Books.
Jarvie, I. (2010) 'Dissent about descent: Evolution, design, and education'. *Philosophy of the Social Sciences*, 40: 467–78.
Jay, N. (1992) *Throughout Your Generations Forever: Sacrifice, Religion, and Paternity*. Chicago: University of Chicago Press.
Johnson, P. (1991) *Darwin on Trial*. Downers Grove IL: Intervarsity Press.
Johnson, S. (2008) *The Invention of Air: A Story of Science, Faith, Revolution, and the Birth of America*. New York: Penguin Riverhead.

Johnston, M. (2009) *Surviving Death*. Princeton: Princeton University Press.

Johnstone, H.W., Jr. (1971) 'Some trends in rhetorical theory', in L. Bitzer and E. Black (eds) *The Prospect of Rhetoric*, pp. 77–89. Englewood Cliffs, NJ: Prentice Hall.

Jones, A. (1998) *Science in Faith: A Christian Perspective on Teaching Science*. Bristol UK: Christian Schools' Trust.

Kahn, J. (2004) 'How a drug becomes "ethnic": Law, commerce, and the production of racial categories in medicine'. *Yale J Health Policy Law Ethics*, 4: 1–46.

Kant, I. (1781) *Kritik der reinen Vernunft* ('Critique of pure reason'). Riga, Latvia: Hartknoch.

Kant, I. (1798) *Anthropologie in pragmatischer Hinsicht* ('Anthropology from a pragmatic standpoint'). Königsberg, East Prussia: Nicolovius.

Kay, L. (2000) *Who Wrote the Book of Life? A History of the Genetic Code*. Palo Alto CA: Stanford University Press.

Kent, R. (1981) *A History of British Empirical Sociology*. Aldershot: Gower.

Kevles, D. (1985) *In the Name of Eugenics*. New York: Alfred Knopf.

Khushf, G. (2004) 'A hierarchical architecture for nano-scale science and technology: Taking stock of the claims about science made by advocates of NBIC convergence', in D. Baird, A. Nordmann and J. Schummer (eds) *Discovering the Nanoscale*, pp. 21–33. Amsterdam: IOS Press.

Koerner, L. (1999) *Linnaeus: Nature and Nation*. Cambridge MA: Harvard University Press.

Kostoff, R. (2005) *Systematic Acceleration of Radical Discovery and Innovation in Science and Technology*. DTIC Technical Report Number ADA430720 (Defense Technical Information Center, Fort Belvoir VA). Available from: www.dtic.mil/

Kostoff, R. (2006) 'Structure and infrastructure of global nanotechnology literature'. *Journal of Nanoparticle Research*, 8(3–4): 301–21.

Kostoff, R., S. Bhattacharya and M. Pecht (2007) 'Assessment of China's and India's science and technology literature – introduction, background, and approach'. *Technological Forecasting and Social Change*, 74: 1519–38.

Kuhn, T.S. (1970) *The Structure of Scientific Revolutions*. 2nd edn (Orig. 1962). Chicago: University of Chicago Press.

Kuhn, T.S. (1977) 'Comment on the relations of science and art', in *The Essential Tension*, pp. 340–52. Chicago: University of Chicago Press.

Kurzweil, R. (1999) *The Age of Spiritual Machines*. New York: Random House.

Kurzweil, R. (2005) *The Singularity is Near: When Humans Transcend Biology*. New York: Viking.

Kurzweil, R. (2006) 'Nanotechnology dangers and defences'. *Nanotechnology Perceptions*, 2: 7–13.

Langholm, O. (1998) *The Legacy of Scholasticism in Economic Thought*. Cambridge UK: Cambridge University Press.

Latour, B. (1987) *Science in Action*. Milton Keynes UK: Open University Press.

Latour, B. (1993) *We Have Never Been Modern*. Cambridge MA: Harvard University Press.

Latour, B. (2004) *Politics of Nature*. Cambridge MA: Harvard University Press.

Laudan, L. (1981a) 'A problem-solving approach to scientific progress', in I. Hacking (ed.) *Scientific Revolutions*, pp. 144–55. Oxford: Oxford University Press.

Laudan, L. (1981b) *Science and Hypothesis*. Dordrecht: Kluwer.
Lenin, V.I. (1948) *Imperialism, the Highest Stage of Capitalism* (Orig. 1917). London: Lawrence and Wishart.
Lenski, R., C. Ofria, R. Pennock and C. Adami (2003) 'The evolutionary origin of complex features'. *Nature*, 423: 139–44.
Lepenies, W. (1988) *Between Literature and Science: The Rise of Sociology*. Cambridge UK: Cambridge University Press.
Levine, D. (1995) *Visions of the Sociological Tradition*. Chicago: University of Chicago Press.
LUDBS (Lehigh University Department of Biological Sciences) (2007) Department Position on Evolution and 'Intelligent Design'. http://www.lehigh.edu/bio/news/evolution.htm.
Lukes, S. (1996) *The Curious Enlightenment of Professor Caritat: A Comedy of Ideas*. London: Verso.
MacIntyre, A. (1981) *After Virtue*. South Bend IN: Notre Dame Press.
MacIntyre, A. (1999) *Dependent Rational Animals*. La Salle IL: Open Court Press.
Masuzawa, T. (2005) *The Invention of World Religions*. Chicago: University of Chicago Press.
Maxwell, N. (2007) *From Knowledge to Wisdom*. 2nd edn (Orig. 1984). London: Pentire Press.
Mayr, E. (1970) *Populations, Species and Evolution*. Cambridge MA: Harvard University Press.
Mayr, E. (1982) *The Growth of Biological Thought*. Cambridge MA: Harvard University Press.
Mayr, E. (1994) 'The advance of science and scientific revolutions'. *Journal of the History of the Behavioral Sciences*, 30: 328–34.
Mayr, E. (2002) 'The biology of race and the concept of equality'. *Daedalus*, Winter, pp. 89–94.
Mazlish, B. (1993) *The Fourth Discontinuity: The Co-Evolution of Humans and Machines*. New Haven CT: Yale University Press.
McEvoy, J. (2010) *The Historiography of the Chemical Revolution*. London: Pickering and Chatto.
McLuhan, M. (1964) *Understanding Media*. New York: McGraw-Hill.
McVeigh, K. (2009) 'Judge rules activist's beliefs on climate change akin to religion'. *Guardian* (London), 3 November.
Meller, H. (1994) *Patrick Geddes: Social Evolutionist and City Planner*. London: Routledge.
Merton, R.K. (1970) *Science, Technology and Society in Seventeenth Century England*. New York: Harper & Row.
Meyer, S.C. (2009) *Signature in the Cell*. New York: HarperOne.
Michael, R. (1990) *Bionomics: The Inevitability of Capitalism*. New York: Henry Holt.
Michels, R. (1959[1915]) *Political Parties*. New York: Dover.
Milbank, J. (1990) *Theology and Social Theory*. Oxford: Blackwell.
Milmo, C. (2007) 'Fury at DNA pioneer's theory: Africans are less intelligent than Westerners'. *The Independent*, London, 17 October.
Mirowski, P. (2002) *Machine Dreams: Economics Becomes a Cyborg Science*. Cambridge UK: Cambridge University Press.
Molnar, T. (1972) *Utopia: The Perennial Heresy*. London: Tom Stacey.
Moore, B. (1970) *Reflections on the Causes of Human Misery*. Boston: Beacon Press.

Morange, M. (1998) *A History of Molecular Biology*. Cambridge MA: Harvard University Press.
More, M. (2005) 'The proactionary principle'. http://www.maxmore.com/proactionary.htm.
Morris, S. (2008) 'Second life affair leads to real life divorce'. *Guardian* (London), 13 November.
Morris, S.C. (2003) *Life's Solution: Inevitable Humans in a Lonely Universe*. Cambridge UK: Cambridge University Press.
Moss, L. (2003) *What Genes Can't Do*. Cambridge MA: MIT.
Murray, C. (2003) *Human Accomplishment*. New York: HarperCollins.
Nadler, S. (2008) *The Best of All Possible Worlds*. New York: Farrar Straus Giroux.
Nagel, T. (1986) *The View from Nowhere*. Oxford: Oxford University Press.
Needham, J. (1976) 'The evolution of oecumenical science: The roles of Europe and China'. *Interdisciplinary Science Reviews*, 1(3): 202–14.
Nisbett, R. (2002) *The Geography of Thought: How Asians and Westerners Think Differently – and Why*. London: Nicholas Brealey.
Noble, D.F. (1997) *The Religion of Technology: The Spirit of Invention and the Divinity of Man*. New York: Penguin.
Nordmann, A. (ed.) (2004) *Converging Technologies – Shaping the Future of Human Societies*. Brussels: European Commission.
Norman, D. (1993) *Things That Make Us Smart*. Reading MA: Addison-Wesley.
Nussbaum, M. (1986) *The Fragility of Goodness*. Cambridge UK: Cambridge University Press.
Nutt, D. (2009) 'Equasy – An overlooked addiction with implications for the current debate on drug harms'. *Journal of Psychopharmacology*, 23: 3–5.
Offer, J. (1999) 'Spencer's future of welfare: A vision eclipsed', *Sociological Review*, 47: 136–62.
Parikka, J. (2008) 'Politics of swarms: Translations between entomology and biopolitics'. *Parallax*, 14(3): 112–24.
Parsons, T. (1937) *The Structure of Social Action*. New York: Harper and Row.
Passmore, J. (1966) *A Hundred Years of Philosophy*. Harmondsworth UK: Penguin.
Passmore, J. (1970) *The Perfectibility of Man*. London: Duckworth.
Paul, D. (1998) *The Politics of Heredity*. Albany NY: SUNY Press.
Paul, D. (2005) 'Genetic engineering and eugenics: The uses of history', in H.W. Baillie and T.K. Casey (eds) *Is Human Nature Obsolete?*, pp. 123–52. Cambridge MA: MIT Press.
Peart, S. and D. Levy (2008) 'Darwin's unpublished letter at the Bradlaugh–Besant trial: A question of divided expert judgment'. *European Journal of Political Economy* 24: 343–53.
Pennock, R. (2005) 'Expert testimony'. *Kitzmiller v. Dover Area School District*, Transcript Day 3, 28 September.
Pennock, R. (2010) 'Can't philosophers tell the difference between science and religion?: Demarcation revisited'. *Synthese*, 172, forthcoming.
Piccone, P. and G. Ulmen (2002) 'The uses and abuses of Carl Schmitt'. *Telos*, 122: 3–32.
Pichot, A. (2009) *The Pure Society: From Darwin to Hitler*. London: Verso.
Pigou, A. (2000[1920]) *The Economics of Welfare*. New Brunswick NJ: Transaction.
Polanyi, M. (1957) *Personal Knowledge*. Chicago: University of Chicago Press.
Popper, K. (1946) *The Open Society and Its Enemies*. New York: Harper and Row.

Popper, K. (1957) *The Poverty of Historicism*. New York: Harper and Row.
Popper, K. (1963) *Conjectures and Refutations*. New York: Harper and Row.
Por, F.D. (2006) 'The actuality of Lamarck: Towards the bicentenary of his *Philosophie Zoologique*'. *Integrative Biology*, 1: 48–52.
Porter, A., J. Youtie, P. Shapira, D. Schoeneck (2008) 'Refining search terms for nanotechnology'. *Journal of Nanoparticle Research*, 10(5): 715–28.
Porter, T. (1986) *The Rise of Statistical Thinking, 1820–1900*. Princeton: Princeton University Press.
Postman, N. (1985) *Amusing Ourselves to Death: Public Discourse in the Age of Show Business*. New York: Penguin.
Price, C. (1993) *Time, Discounting and Value*. Oxford: Blackwell.
Proctor, R. (1988) *Racial Hygiene: Medicine under the Nazis*. Cambridge MA: Harvard University Press.
Proctor, R. (1991) *Value-Free Science? Purity and Power in Modern Science*. Cambridge MA: Harvard University Press.
Pyle, A. (2003) *Malebranche*. London: Routledge.
Rasmussen, N. (1997) 'The mid-century biophysics bubble: Hiroshima and the biological revolution in America, revisited'. *History of Science*, 35: 245–91.
Rawls, J. (1971) *A Theory of Justice*. Cambridge MA: Harvard University Press.
Rayward, W.B. (1999) 'H.G. Wells's idea of a world brain: A critical re-assessment'. *Journal of the American Society for Information Science*, 50 (May 15): 19–22.
Renwick, C. (2011) 'Observation and detachment: William Beveridge and "The National Bases of Social Science"'. *History of Political Economy*, 43 (Supplement).
Renwick, C. (2012) *British Sociology's Lost Biological Roots*. London: Palgrave Macmillan.
Rescher, N. (1984) *The Limits of Science*. Berkeley: University of California Press.
Richards, J. (ed.) (2002) *Are We Spiritual Machines?* Seattle: Discovery Institute.
Richards, R.J. (2002) 'The linguistic creation of man: Charles Darwin, August Schleicher, Ernst Haeckel, and the Missing Link in 19th-century evolutionary theory', in M. Doerres (ed.) *Experimenting in Tongues: Studies in Science and Language*. Palo Alto, CA: Stanford University Press.
Ripstein, A. (2007) 'As if it never happened'. *William & Mary Law Review*, 48(5): 1957–97.
Rivers, I. and D. Wykes (eds) (2008) *Joseph Priestley: Scientist, Philosopher, and Theologian*. Oxford: Oxford University Press.
Robbins, R.J. (1996) 'Information management as key to human genome project'. http://www.esp.org/rjr/cthsl.pdf
Roco, M.C. (2002) 'Coherence and divergence in megatrends in science and engineering'. *Journal of Nanoparticle Research*, 4: 9–19.
Roco, M.C. and W.S. Bainbridge (eds) (2002a) *Converging Technologies for Improving Human Performance: Nanotechnology, Biotechnology, Information Technology, and Cognitive Science*. NSF/DOC-sponsored report, Arlington VA.
Roco, M.C. and W.S. Bainbridge (2002b) 'Converging technologies for improving human performance: Integrating from the nanoscale'. *Journal of Nanoparticle Research*, 4: 281–95.
Rorty, R. (1965) 'Mind-body identity, privacy, and categories'. *Review of Metaphysics*, 19: 24–54.
Rorty, R. (1998) *Achieving Our Country: Leftist Thought in Twentieth Century America*. Cambridge MA: Harvard University Press.

Rose, N. (2006) *The Politics of Life Itself: Biomedicine, Power, and Subjectivity in the Twenty-First Century*. Princeton NJ: Princeton University Press.
Rosen, M. (2008) 'Robbing the grave of Immanuel Kant'. *Times Literary Supplement*, 15 October.
Rosen, R. (1999) *Essays on Life Itself*. New York: Columbia University Press.
Ross, L. (1977) 'The intuitive psychologist and his shortcomings: Distortions in the attribution process', in L. Berkowitz (ed.) *Advances in Experimental Social Psychology*, vol. 10, pp. 173–220. New York: Academic Press.
Runciman, W.G. (1989) *A Treatise on Social Theory, Vol. II: Substantive Social Theory*. Cambridge UK: Cambridge University Press.
Russel, A. (2009) '"Ilities" – Judging architecture and design'. http://www.uncommondescent.com/intelligent-design/ilities-judging-architecture-and-design/, 24 May.
Said, E. (1978) *Orientalism*. New York: Random House.
Sandel, M. (2007) *The Case against Perfection: Ethics in the Age of Genetic Engineering*. Cambridge MA: Harvard University Press.
Saracci, R. (2001) 'Introducing the history of epidemiology', in J. Orsen et al. (eds) *Teaching Epidemiology*. Oxford: Oxford University Press.
Schaefer, W. (ed.) (1984) *Finalization in Science*. Dordrecht: Kluwer.
Schliesser, E. (2008) 'Hume's Newtonianism and anti-Newtonianism'. *Stanford Encyclopedia of Philosophy*, http://plato.stanford.edu/entries/hume-newton/.
Schmidt, J.C. (2007) 'Knowledge politics of interdisciplinarity: Specifying the type of interdisciplinarity in the NSF's NBIC scenario'. *Innovation*, 20(4): 313–27.
Schmitt, C. (1996) *The Concept of the Political* (Orig. 1932). Chicago: University of Chicago Press.
Schnädelbach, H. (1984) *Philosophy in Germany, 1831–1933*. Cambridge UK: Cambridge University Press.
Schneewind, J. (1984) 'The divine corporation and the history of ethics', in R. Rorty, J. Schneewind and Q. Skinner (eds) *Philosophy in History*, pp. 173–92. Cambridge UK: Cambridge University Press.
Schneewind, J. (1997) *The Idea of Autonomy*. Cambridge UK: Cambridge University Press.
Schofield, R. (1997) *The Enlightenment of Joseph Priestley, 1733–1773*. University Park PA: Pennsylvania State University Press.
Schofield, R. (2004) *The Enlightened Joseph Priestley, 1773–1804*. University Park PA: Pennsylvania State University Press.
Schrödinger, E. (1955) *What is Life? The Physical Aspects of the Living Cell* (Orig. 1944). Cambridge UK: Cambridge University Press.
Schumpeter, J. (1961) *The Theory of Economic Development* (Orig. 1912). New York: Galaxy Books.
Selin, C. (2007) 'Expectations and the emergence of nanotechnology'. *Science, Technology and Human Values*, 32: 196–220.
Shapin, S. and S. Schaffer (1985) *Leviathan and the Air-Pump*. Princeton: Princeton University Press.
Shearmur, J. (2010) 'Steve Fuller and intelligent design'. *Philosophy of the Social Sciences*, 40: 433–45.
Shew, A. (2008) 'Nanotechnology's future: Considerations for the professional', in F. Allhoff and P. Lin (eds) *Nanotechnology and Society: Current and Emerging Debates*. Dordrecht: Springer.

Simon, H. (1977) *The Sciences of the Artificial*. 2nd edn. Cambridge MA: MIT Press.
Singer, P. (1975) *Animal Liberation*. New York: Random House.
Singer, P. (1993) *Practical Ethics*. Cambridge UK: Cambridge University Press.
Singer, P. (1994) *Rethinking Life and Death*. Melbourne: Text Publishing Company.
Singer, P. (1999) *A Darwinian Left*. London: Weidenfeld & Nicolson.
Singer, P. (2004) *One World: The Ethics of Globalization*. New Haven CT: Yale University Press.
Slovic, P. (2007) 'Genocide: When compassion fails'. *New Scientist*, 2598 (7 April): 18.
Smith, R. (2007) *Being Human: Historical Knowledge and the Creation of Human Nature*. New York: Columbia University Press.
Snyder, L. (2006) *Reforming Philosophy: A Victorian Debate on Science and Society*. Chicago: University of Chicago Press.
Sober, E. (2008) *Evidence and Evolution*. Cambridge UK: Cambridge University Press.
Sorokin, P. (1928) *Contemporary Sociological Theories*. New York: Harper and Row.
Stark, R. and W. Bainbridge (1996) *A Theory of Religion* (Orig. 1987). New Brunswick NJ: Rutgers University Press.
Stehr, N. and H. von Storch (2005) 'Editorial: Introduction to papers on mitigation and adaptation strategies for climate change: Protecting nature from society or protecting society from nature?', *Environmental Science and Policy*, 8: 537–40.
Stehr, N. and H. von Storch (eds) (2000) *The Sources and Consequences of Climate Change and Climate Variability in Historical Times*. Dordrecht: Kluwer.
Stephenson, N. (1992) *Snow Crash*. New York: Bantam.
Sterelny, K. (2001) *Dawkins vs. Gould: Survival of the Fittest*. Cambridge UK: Icon.
Stix, G. (1996) 'Trends in nanotechnology: Waiting for breakthroughs'. *Scientific American* (April): 94–9.
Strauss, L. (2000) *On Tyranny* (Orig. 1948). Chicago: University of Chicago Press.
Stueber, K. (2006) *Rediscovering Empathy: Agency, Folk Psychology, and the Human Sciences*. Cambridge MA: MIT Press.
Talbot, M. (2009) 'Can a daily pill really boost your brain power?'. *Observer* (London), 20 November.
Täljedal, I.-B. (2010) 'Linnæus's Questions to the Sami Herdsman: The ID-versus-science issue in a nutshell'. *Philosophy of the Social Sciences*, 40: 456–66.
Teilhard de Chardin, P. (1961) *The Phenomenon of Man* (Orig. 1955). New York: Harper and Row.
Tierney, J. (2009) 'Guilt and atonement on the path to adulthood'. *The New York Times*, 24 August.
Toulmin, S. (1961) *Foresight and Understanding*. Bloomington IN: Indiana University Press.
Toulmin, S. (1982) *The Return of Cosmology: Postmodern Science and the Theology of Nature*. Berkeley: University of California Press.
Toulmin, S. (1990) *Cosmopolis: The Hidden Agenda of Modernity*. New York: Free Press.
Turkle, S. (1984) *Second Self: Computers and the Human Spirit*. New York: Simon and Schuster.
Turner, S. (2002) *Brains/Practices/Relativism: Social Theory after Cognitive Science*. London: Sage.

Turner, S. (2007) 'Social theory as cognitive neuroscience'. *European Journal of Social Theory*, 10: 357–74.
Urry, J. (2000) *Sociology Beyond Societies*. London: Routledge.
US District Court. (2005) *Kitzmiller v. Dover Area School District*. Middle District of Pennsylvania: Harrisburg PA, USA. www.pamd.uscourts.org/kitzmiller
Voegelin, E. (1952) *The New Science of Politics*. Chicago: University of Chicago Press.
Wallerstein, I. (ed.) (1996) *Open the Social Sciences*. Palo Alto: Stanford University Press.
Watson, J.D. (2003) *DNA: The Secret of Life*. New York: Alfred Knopf.
Weber, B. and D. Depew (eds) (2003) *Evolution and Learning: The Baldwin Effect Reconsidered*. Cambridge MA: MIT Press.
Weber, M. (1963) *The Sociology of Religion*. Trans. E. Fischoff (Orig. 1922). Boston: Beacon Press.
Weigmann, K. (2006) 'Racial medicine: Here to stay?'. *EMBO Reports*, 7: 246–9.
Weikart, R. (2005) *From Darwin to Hitler*. New York: Macmillan.
Werskey, G. (1978) *The Visible College: A Collective Biography of British Scientists and Socialists of the 1930s*. London: Free Association Press.
Wiener, N. (1964) *God and Golem, Inc*. Cambridge MA: MIT Press.
Williams, B. (1973) *Problems of the Self*. Oxford: Oxford University Press.
Williams, T. (2007) 'John Duns Scotus', in E. Zalta (ed.) *Stanford Encyclopedia of Philosophy*, http://plato.stanford.edu/entries/duns-scotus/.
Wilmut, I., K. Campbell, C. Tudge (2000) *The Second Creation: Dolly and the Age of Biological Control*. New York: Farrar, Straus and Giroux.
Wilson, C. (2004) *Moral Animals: Ideals and Constraints*. Oxford: Oxford University Press.
Wilson, E.O. (1975) *Sociobiology: The New Synthesis*. Cambridge MA: Harvard University Press.
Wilson, E.O. (1998) *Consilience: The Unity of Knowledge*. New York: Alfred Knopf.
Wise, M.N. (1989) 'Work and waste: Political economy and natural philosophy in nineteenth century Britain'. *History of Science*, 27: 263–301.
Woese, C. (1998) 'The universal ancestor'. *Proceedings of the National Academy of Sciences*, 95 (June): 6854–9.
Woese, C. (2004) 'A new biology for a new century'. *Microbiology and Molecular Biology Reviews*, 68: 173–86.
Wolbring, G. (2006) 'Ableism and NBICS', http://www.innovationwatch.com/choiceisyours/choiceisyours.2006.08.15.htm, 15 August.
Wood, R. and V. Orel (2005) 'Scientific breeding in central Europe in the early nineteenth century: Background to Mendel's later work', *Journal of the History of Biology*, 38: 239–72.
Woodward, T. (2003) *Doubts about Darwin: A History of Intelligent Design*. Grand Rapids MI: Baker.
Woolgar, S. (1991) 'Configuring the user: The case of usability trials', in J. Law (ed.) *A Sociology of Monsters: Essays on Power, Technology and Domination*, pp. 58–97. London: Routledge.
Yang, A. (2008) 'Matters of demarcation: Philosophy, biology, and the evolving fraternity between disciplines'. *International Studies in the Philosophy of Science*, 22: 211–25.
Young, B. (2000) 'Malthus among the Theologians'. *Clio Medica*, 59: 93–113.

Youtie, J., M. Iacopetta and S. Graham (2008) 'Assessing the nature of nanotechnology: Can we uncover an emerging general purpose technology?'. *Journal of Technology Transfer*, 33: 315–29.

Zeigarnik, B. (1967) 'On finished and unfinished tasks' (Orig. 1927), in W.D. Ellis (ed.) A *Sourcebook of Gestalt Psychology*. New York: Humanities Press.

Zimmerman, A. (2001) *Anthropology and Antihumanism in Imperial Germany*. Chicago: University of Chicago Press.

Index